Springer Series in Biophysics 4

Editor: P.M. Bayley, London

Springer Series in Biophysics

Volume 1 **Structure, Dynamics and Function of Biomolecules**
Edited by A. Ehrenberg, R. Rigler, A. Gräslund and L. Nilsson (1987)

Volume 2 **Biophysics and Synchrotron Radiation.** Edited by A. Bianconi and A. Congiu Castellano (1987)

Volume 3 **Cytoskeletal and Extracellular Proteins**
Edited by U. Aebi and J. Engel (1989)

Volume 4 **Electron Probe Microanalysis**
Edited by K. Zierold and H.K. Hagler (1989)

K. Zierold H.K. Hagler (Eds.)

Electron Probe Microanalysis

Applications in Biology and Medicine

With 84 Figures

Springer-Verlag Berlin Heidelberg New York
London Paris Tokyo Hong Kong

Dr. KARL ZIEROLD
Max-Planck-Institut für Systemphysiologie
Rheinlanddamm 201
4600 Dortmund 1, FRG

HERBERT K. HAGLER, Ph.D.
Associate Professor of Pathology
Department of Pathology
University of Texas
Southwestern Medical Center
5323 Harry Hines Blvd.
Dallas, TX 75235-9072, USA

Inivited contributions with conference discussion of the conference on »Progress of Electron Probe Microanalysis in Biology and Medicine« at Schloss Ringberg, November 16-19, 1988.

Library of Congress Cataloging-in-Publication Data. Conference on Progress of Electron Probe Microanalysis in Biology and Medicine (1988 : Schloss Ringberg) Electron probe microanalysis : applications in biology and medicine : invited contributions with conference discussion of the conference on "progress of electron probe microanalysis in biology and medicine" at Schloss Ringberg, November 16–19, 1988 / K. Zierold, H.K. Hagler (eds.). p. cm.— (Springer series in biophysics ; v. 4) "Conference on Progress of Electron Probe Microanalysis in Biology and Medicine, Schloss Ringberg, November 16–19, 1988"—T.p. verso. Includes index.

ISBN-13: 978-3-642-74479-2 e-ISBN-13: 978-3-642-74477-8
DOI: 10.1007/ 978-3-642-74477-8

1. Electron probe microanalysis—Congresses. I. Zierold, K. (Karl), 1944- . II. Hagler, H.K. (Herbert K.), 1946- . III. Title. IV. Series. [DNLM: 1. Electron Probe Microanalysis—congresses. 2. Microbiological Technics—congresses. 3. Microscopy, Electron—congresses. W1 SP685MD v. 4 / QH 212.E4 C748e 1988] QP519.9.E4333C66 1988 578'.45—dc20 DNLM/DLC for Library of Congress

Softcover reprint of the hardcover 1st edition 1989

2131/3145-543210 – Printed on acid-free paper

Preface

The aim of electron probe microanalysis of biological systems is to identify, localize, and quantify elements, mass, and water in cells and tissues. The method is based on the idea that all electrons and photons emerging from an electron beam irradiated specimen contain information on its structure and composition. In particular, energy spectroscopy of X-rays and electrons after interaction of the electron beam with the specimen is used for this purpose. However, the application of this method in biology and medicine has to overcome three specific problems: 1. The principle constituent of most cell samples is water. Since liquid water is not compatible with vacuum conditions in the electron microscope, specimens have to be prepared without disturbing the other components, in particular diffusible ions (elements). 2. Electron probe microanalysis provides physical data on either dry specimens or fully hydrated, frozen specimens. This data usually has to be converted into quantitative data meaningful to the cell biologist or physiologist. 3. Cells and tissues are not static but dynamic systems. Thus, for example, microanalysis of physiological processes requires sampling techniques which are adapted to address specific biological or medical questions.

During recent years, remarkable progress has been made to overcome these problems. Cryopreparation, image analysis, and electron energy loss spectroscopy are key areas which have solved some problems and offer promise for future improvements. It was the idea of the conference on "electron probe microanalysis in biology and medicine" which we organized at Schloss Ringberg, Germany, on November 16-19, 1988, to present the state of the art and to discuss current problems and future trends. This volume contains the 23 invited contributions to this conference together with the conference discussion based on written questions, answers, and comments. In our opinion this compilation reflects the actual scientific discussion in this field of research. These proceedings and discussions should not be taken as a basic comprehensive textbook on the subject. Well established techniques, for example microanalysis of bulk specimens and microdroplets or microanalytical identification of histochemical reaction products, are omitted in favour of articles on recent techniques and promising applications.

We dedicate this volume to T.A.(Ted) Hall to honour him for his pioneering work for many years in establishing electron probe microanalysis as a useful tool in biological research. He has been a source of inspiration and critical thinking to many of us active in this field of endeavor. The following personal appreciation by B.L. Gupta supports our esteem for him.

The historical overview on electron probe microanalysis in biology should be taken as introduction. Specimen preparation techniques are described and compared by the following three articles. Seven articles concentrate on progress and problems with data acquisition and evaluation in X-ray microanalysis and electron energy loss spectroscopy. The main part (10 articles) of the volume consists of reports on the application of electron probe microanalysis in biological research: Intracellular element localization, studies of epithelial ion transport and of dynamic processes in cells. Applications in medicine, in particular in pathology and diagnosis, are reviewed in the final two articles.

The conference on "electron probe microanalysis in biology and medicine" was financed by Deutsche Forschungsgemeinschaft, Deutsche Gesellschaft für Elektronenmikroskopie, and Max-Planck-Gesellschaft zur Förderung der Wissenschaften e.V. Additional financial support was received by Carl Zeiss (Oberkochen/FRG), Deutsche Gesellschaft für Zellbiologie e.V., Gatan (München/ FRG), Link Analytical (High Wycombe/England),

Outokumpu Electronics (Espoo/Finland), Philips (Eindhoven/Netherlands), Tracor Europa (Bruchsal/FRG). A Macintosh II computer for elemental imaging was provided for the period of the conference by ABS computer (Polling/FRG). We thank them all for their generous support. Finally, we thank the Max-Planck-Gesellschaft for the hospitality at Schloss Ringberg. We are convinced that the friendly atmosphere we have enjoyed there has stimulated discussions and exchange of ideas, and we hope that this volume as a result of this conference will promote further research.

Karl Zierold
Herbert K. Hagler

Dedicated to Theodore Alvin Hall

Theodore Alvin Hall - A Personal Appreciation

Everybody who is interested in biological microprobe analysis of chemical elements knows Ted Hall either through personal contact, by his reputation, or from one of his 160 publications. In the late 1960s, he firmly established that Bremsstrahlung, "white" or "continuum" X-ray radiations from thin films (= sections in biology) are directly proportional to the total mass per unit area of the specimen under the electron beam. The publication "Quantitative microprobe analysis of thin specimens: Continuum method" (Hall & Werba, 1971) hailed the dawn of fully quantitative X-ray microanalysis in biology. Hall published a full treatment of this approach in his 120 page chapter on "The microprobe assay of chemical elements" (In "Physical Techniques in Biological Research" 2nd Edition Vol. 1A (ed. G. Oster) Academic Press, New York), which has since served as a primary reference work on the subject. His short paper in 1973 on "The use of thin specimens for X-ray microanalysis in biology" (J. Microscopy 99, 177-182) is probably the most frequently cited single reference on the quantification of X-ray data. Not only is Ted Hall revered in biological microprobe laboratories around the world but his commercial value can be gauged by the fact that all manufacturers of "inclusive" spectroscopic equipment claim to use the "Hall method" as the hallmark of reliability for their quantification soft-ware packet, even though Ted Hall has been directly involved in the development of only the Link-Systems QUANTEM-FLS software.

In spite of Ted's fundamental contributions to science, he is rather a modest person who projects a low profile of himself and his work. He never gets too excited about any results and wants to know what they actually mean. Ted's measured caution and foresight in accurately gauging the magnitude of a scientific problem is exemplified by an amusing episode. Ted Hall immigrated in 1962 from Cornell University Medical School, New York to the Cavendish' Laboratory in Cambridge to join the microprobe group of Dr. Cosslett's world renowned electron microscopy laboratory. Ted Hall, although a physicist, was primarily interested in applying the X-ray microprobe analysis to biomedical specimens. In the Zoology Department nearby there was at that time a great interest in methods for measuring ions and water in transporting epithelia such as insect Malpighian tubules. The late Ken Machin, a bright physicist, working as the resident biophysicist in Zoology, was despatched to Cavendish to see this "young American" and find out if his new method of X-ray microanalysis could be used for the purpose. Ted listened to Ken's precise description of the problem and then told him to come back in ten years for the answer. This prediction made in 1963 turned out to be remarkably accurate. In 1974 in the newly established Biological Microprobe Laboratory in the Zoology Department, Ted Hall and I used 1-μm thick, frozen-hydrated sections from the Malpighian tubules of the blowfly, Calliphora erythrocephala, JEOL JXA-50A, 2 X-ray wavelength dispersive spectrometers (WDS) with Ted's home-made proportional counter to measure continuum, and reported the first fully quantitative data obtained in situ for the distribution of Na and K.

Ted Hall was born in 1925 in Far Rockaway in Queens, New York. He graduated in 1944 in Physics from Harvard University and obtained a Ph.D. in 1950 in Nuclear Physics from the University of Chicago. After a sojourn with nuclear physics, and distressed by its use for developing the atomic bomb, Ted became interested in the application of physical techniques to biology and in 1952 moved to the Memorial Centre for Cancer, Cornell University Medical School. One of the first challenges he faced was to localize zinc in the prostrate glands and spermatozoa. Since it is not convenient (or humane) to perform coitus interruptus on laboratory rats for obtaining sperm ejaculates, he designed "An electro-ejaculation technique for rats" (Anat Rec 140, 49-50). The rumour has it that a genetically engineered biological-chip version of this apparatus now forms a part of the female genitalia of rats all over the world! It also shows that unlike Sergei Eisenstein, the Russian movie genius, Ted Hall has a perfectly normal interest in heterosexual biology.

From 1971 to 1976 Ted wore, to twist a phrase, two different shoes. (I have never known Ted to wear a hat anyway!). He became one of the five Principal Participants in the Biological Microprobe Laboratory in the Zoology Department, primarily responsible for the analytical instrumentation to be used for 1-µm thick frozen-hydrated cryosections. At the same time, he was also awarded a grant to set up an EMMA-4 in the Cavendish Laboratory for analysing ultrathin biological sections. In the "EMMA Lab" Ted and his collaborator Patricia Peters worked with numerous biologists from Cambridge, Britain and abroad mainly to localize calcium in a variety of biological materials. Although this EMMA work was never fully quantitative, nor generally extended to cryosections, it made an important contribution in establishing the role of Ca^{2+} as the second messenger in cells. The list of biologists who came to work with Ted in his two microprobe laboratories in Cambridge reads like "who is who" in biological microanalysis. It was on EMMA-4 where Andy and Avril Somlyo came in 1974 to sit at the feet of the Guru, and be initiated into the mysteries of X-ray microanalysis.

On the exodus of the Cavendish Laboratory to the outskirts of Cambridge, Ted Hall had to move to Zoology in 1976 until in 1984 he was forced to take an early retirement because of a complete lack of research funds. (The Biological Microprobe Lab. in Zoology has now completely disappeared and all the instruments have been sold). The proceedings of the present symposium show that since the pioneering work of Ted Hall, biological microprobe analysis has come of age. It is a great pity that Ted Hall's unique experience and understanding of the analytical problems can no longer be practically deployed to resolve new technical challenges in this rapidly expanding field. Nevertheless, we all wish that he spends his retirement in happy, fruitful, and enjoyable pursuits!

Brij L. Gupta
Cambridge University
England

Contents

List of Authors

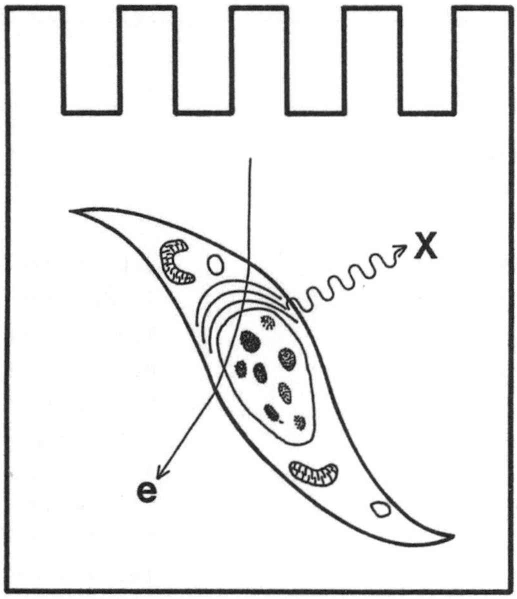
X
e

The History of Electron Probe Microanalysis in Biology

T. A. Hall
49 Owlstone Road
Cambridge CB3 9JH
England

Introduction

A systematic discussion of biological electron-probe microanalysis would examine one-by-one several aspects of the subject, including (at least) specimen preparation, specimen imaging, analytical spatial resolution, concentration limits for elemental detection, beam damage, methods of quantitative analysis and a review of the categories of biological studies and of the notable biological results (for example, see Hall, 1979.) Instead, here we shall try to trace the actual path of development of the method, a development shaped of course by the interaction of all of these aspects as the available instrumentation became ever more effective.

I shall try to describe <u>what</u> happened <u>when</u>, but my concern is *not* with <u>who</u>. Common interests and common instrumentation led to parallel lines of development in many laboratories. This historical sketch is based to a considerable extent on the material I know best - my own experience - but no claims of priority are implied.

Pre-history

I believe that before the advent of the electron-probe microanalyser, elemental microanalysis did not really exist. There were several techniques capable of detecting elements at very low (so-called "trace") concentrations, notably mass spectrometry, optical emission spectrometry, atomic absorption spectrometry and neutron activation analysis, but there was no way to determine the concentration of a given element *in situ* within, say, a single cell (see Hall, 1961). Radioautography reflected turnover rates rather than concentrations, and microprecipitation methods did not give reliable localisations or elemental specificity; and neither of these two methods gave a decent coverage of the periodic table. Among existing techniques, the one coming closest to the goal of *in situ* quantitative microanalysis was perhaps absorption microspectrometry, where an optical microscope is used to determine the optical absorption characteristics of a selected microarea within a tissue section. Irradiation with microbeams of ultra-violet light near the absorption peaks of nucleic acids and proteins could then provide maps of the distributions of these substances. But there was simply no way to gain similar information about the chemical elements as such.

Early biological work with the electron probe (around 1961).

So far as I am aware, the earliest biological microprobe studies were done with two home-made instruments in Washington D.C. (Brooks *et al.*, 1962; Tousimis and Adler, 1963) and with a home-made device in Cambridge, UK (Lever and Duncumb, 1961; Robertson *et al.*, 1961). The instruments in Washington had no facilities for scanning the electron beam; they relied on built-in optical microscopes and on movement of the specimen stage for viewing the specimen and locating the microprobe. The Cambridge studies used the first scanning microanalyser, built by Duncumb out of an RCA electron microscope (Duncumb, 1959). The availability of backscattered-electron or secondary-electron scanning images meant that a static probe could be readily positioned within the specimen by reference to the scanning electron image (it was not always appreciated that the scanning image and the apparent position of the static probe might be out of register!). The main impetus for all of these instruments came from the materials sciences, so specimens tended to be in bulk form and to be mounted on stubs or glass slides. The X-ray spectrometers were all based on curved diffracting crystals (the SiLi energy-dispersive spectrometers were not to enter the biological scene until about 1970).

As demonstrated in the four biological studies cited above, under these conditions one could investigate major elements, like calcium in developing bone, and a variety of abnormal elemental "hot spots" arising from intoxication or disease, but the techniques were inadequate for the study of most of the important elements in normal soft tissues.

Main developments over the next decade

Two developments in the early 1960's greatly increased the effectiveness of electron-probe X-ray microanalysis (EPXMA) in biology. One was the introduction of fully focussing linear diffracting-crystal spectrometers (see Wittry, 1969). The consequent improvement in X-ray collecting power and in peak/background ratios enabled the microanalysis of many important elements (Na, Mg, Si, P, S, Cl, K, Ca, Fe) in normal tissues, even with moderate probe currents (≃ 10 nA). Further, the analysis of microvolumes of body fluids became practical (see Lechene, 1974; Roinel and de Rouffignac, 1982).

The second development, trivial in execution but major in consequence, was the study of thin specimens, especially tissue sections, mounted on ultrathin supporting films. This configuration could provide greatly improved analytical spatial resolution, better than 1 μm, in soft-tissue specimens, while avoiding the backgrounds generated when a microbeam impinges on a bulk support. Of course the X-ray counting rate obtained from a thin specimen is roughly proportional to specimen thickness and is much less than the rate from a thick specimen. In my own laboratory, with a diffracting spectrometer of very modest performance, in order to get adequate counting rates one had to use probe currents ranging up to 1 μA. For many months, our thin specimens promptly disintegrated when exposed to such currents, but then we learned to coat them heavily with evaporated aluminium, about 30 nm applied to both top and bottom faces. With such specimens we did not see any beam damage, so we innocently assumed that there was no beam damage (Hall *et al.*, 1966).

With respect to analytical spatial resolution, Andersen (1967a,b) showed that in bulk specimens, one could expect to achieve resolution of a few μm by operating at low beam voltage to restrict the range of the incident electrons. (Indeed, Swift (1979) later achieved a resolution of ≈ 0.1 μm for the analysis of sulphur in the layers of human hair, but only by using a beam voltage so low that most other elements were not observable.) For a thin specimen mounted on an ultrathin support, confusion about optimum operating conditions persisted for some time, especially among analysts accustomed to bulk specimens, but theory showed that from every major standpoint (resolution, sensitivity and beam damage), high beam voltages were best (Hall, 1968).

The quantitative analysis of bulk specimens posed no fundamental problems -- the formulations already developed in the materials sciences were applicable, more or less -- but for thin specimens the situation was quite different. Since counting rates in a thin specimen are proportional to local elemental mass/area (virtually by definition of "thin"), elemental *ratios* could be obtained readily from the simple equation

$$C_1/C_2 = k\ n_1/n_2 \qquad (1)$$

where C_1 is the concentration of element 1; n_1 is the number of X-ray counts for element 1; and the constant k is obtained with any thin standard containing the two elements in known proportion. But absolute measurements of concentration posed a new problem, as there was no way to know if local variations in an element's X-ray count rate reflected actual variation in local mass-fraction or simply local variation in "mass thickness".

One solution to this problem was to normalise the observed characteristic X-ray count against the continuum count taken in a peak-free region of the spectrum. Originally this procedure was regarded as a gross correction "to avoid being badly misled" (Hall, 1968, p. 272), but it soon developed into a complete quantitative formalism (Hall *et al.*, 1966; Marshall and Hall, 1966, 1968; Hall and Werba, 1971; Hall, 1971). The underlying equation, taking account of the dependence of continuum intensity on atomic number, is

$$n_x/n_b = k\ N_x/\Sigma(NZ^2) \qquad (2)$$

where n_x is the characteristic X-ray signal for element x; n_b is the continuum signal; N_x is the number of atoms of x per unit volume; Z is atomic number; and the sum is taken over all constituent elements. The factor k is obtained from any thin standard of known composition.

The nature of quantitative measurements in thin specimens was discussed by Höhling and Hall (1969). The intensity of a characteristic X ray is proportional to local elemental mass/area, and hence is proportional to mmol of element per unit volume in tissue sections of uniform thickness; the ratio of characteristic and continuum signals is always proportional to mass-fraction, *i.e.* mmol of element per unit specimen mass. Thus for example, using both measures in freeze-dried sections of skin, Sims and Hall (1968) could show that in the process of keratinisation, sulphur mass-fractions do not rise much but sulphur/volume increases as the tissue becomes denser (as the tissue dry-mass fraction rises).

With the technical developments of the 1960's, microprobe analysis became useful to biology. A rather haphazard compilation of publications (Hall, 1971, pp. 266) listed 12 microprobe studies of the fate of foreign bodies in organisms, 16 on the fate of injected or ingested substances, 20 on metabolic abnormalities, 34 on normal physiology (of soft tissues), 12 on prestages of mineralization, 35 on mineralized tissues (mainly bone and teeth), and 8 on the performance of histological stains. (For a contemporary evaluation of these studies, see Hall *et al.*, 1972.)

For most of these studies, specimens were prepared by procedures well established in optical and electron microscopy (fixation, staining, dehydration, embedding). One tended not to worry much that the elements of interest might be removed or displaced during preparation.

Post-1970

In the 1970's striking experiences compelled recognition of the really obvious dangers in conventional specimen preparation. I recall three episodes in my own laboratory:

1. An early effort by Andrew and Avril Somlyo to demonstrate the uptake of barium, as a calcium analogue, into muscle-cell mitochondria. In sections stained with uranyl acetate, barium was found nowhere and the observed mitochondrial "hot-spots" consisted predominantly of uranium; barium was found in the mitochondria only when the staining step was omitted (see Somlyo *et al.*, 1974.)
2. An effort by Berridge, Oschman and Wall to demonstrate a high concentration of calcium in *Calliphora* salivary-gland mitochondria by trapping the calcium through co-precipitation with lead. The observed calcium hot-spots were shown to be artefactual, a result of the diffusion of calcium into the mitochondria during specimen preparation. (Recent microprobe data, Somlyo *et al.* 1985, have shown that calcium concentrations in mitochondria are actually typically in the neighbourhood of 1 mmol calcium/kg dry tissue.)
3. A study of the concentrations of sodium and other electrolytes in the nuclei of chick-embryo erythrocytes. This work was undertaken in the expectation of confirming the then-fashionable belief, based on dubious preparations analysed by other techniques, that sodium levels in cell nuclei are often similar to the levels in extracellular fluid, but the *in situ* analyses of frozen-dried cryosections proved that belief to be false (Jones *et al.*, 1979). (For a survey of microprobe measurements on cryosections, comparing nuclei and cytoplasm, see Gupta and Hall, 1982, Table 3.)

Today, if there is any question of inadvertent displacement of the elements of interest, the only respectable course is cryopreparation, starting with the quench-freeze of a small sample. After this step the alternative routes are: a) analysis of the bulk sample, either directly in the frozen-hydrated state (Lechene *et al.*, 1979; Forrest and Marshall, 1976; Marshall, 1982; Echlin and Taylor, 1986; Zs.-Nagy *et al.*, 1982) or after freeze-drying (Zs.-Nagy *et al.*, 1981); b) cryosectioning and freeze-drying of the section; c) cryosectioning and direct analysis of the frozen-hydrated section; or d) freeze-substitution or freeze-drying of the entire sample, followed by embedding and possibly sectioning.

Probably the most generally applicable of these routes is the analysis of frozen-dried cryosections, which goes back to 1972 at least (Appleton, 1972). However the only way to get reliable data on the composition of matrix-free extracellular spaces in tissues is the analysis of frozen-hydrated material. On the basis of experience with ultrathin sections, the analysis of frozen-hydrated sections has often been deemed impossible, but the successful analysis of 1-μm frozen-hydrated cryosections goes back to 1974 (Gupta, 1976), and later results put the feasibility of such analyses beyond doubt (Gupta *et al.*, 1978; Gupta and Hall, 1981). The remaining alternative, the analysis of cryo-quenched embedded material, was introduced as early as 1967 (see Ingram and Hogben, 1968) and there is now a revival of interest in it (J. and R. Wroblewski, 1986). This route can provide good morphological preservation, but fully quantitative analysis is precluded by the presence of embedding medium, and there are some doubts about possible elemental displacements.

In the 1970's, two important technical developments matured synergistically. First was the construction of instruments combining transmission electron microscopy with X-ray microanalysis ("EMMA"s). Given images of TEM quality, EMMA operators could locate their probes with much greater refinement -- for instance, within or excluding mitochondria-- even if the analytical spatial resolution could not match the image resolution.

The second big development was the solid-state (SiLi) energy-dispersive X-ray detector, which was applied to biological specimens near the end of the 60's (for example, Russ and McNatt, 1969).

In the prototype EMMA (Duncumb, 1966), the X-ray spectrometer originally was a gas-flow proportional counter used in an energy-dispersive mode. This was effective for the analysis of inorganic precipitates but it was quite inadequate for most biology. Extensive biological studies commenced (Höhling *et al.*, 1967) as soon as diffracting-crystal spectrometers were fitted to the column. But the application of diffracting spectrometers to ultrathin specimens necessitated high probe currents (typically 100 nA) and much beam damage. Hence biological work with the EMMAs did not really "take off" until the 1970's, when the solid-state energy-dispersive spectrometers could be fitted. This led to the modern era, where SiLi-based spectrometers are readily fitted to excellent TEMs, enabling a static-probe analysis for a whole wide range of elements to be completed with a probe current of a few nA and a total local beam exposure of perhaps 100 nanoCoulomb. When operated in the scanning transmission mode, with strongly focussing lenses and bright electron sources, the modern TEMs can focus the necessary probe currents into spot-diameters of the order of 10 nm.

The best performance available from such instrumentation has been fully documented in the long series of papers from the Philadelphia laboratory of Somlyo and Shuman. Among many measures of capabilities, they have demonstrated detectability of calcium down to about 1 mmol Ca/kg even in the presence of normal cell potassium (implicit in Somlyo and Shuman, 1982, plus Shuman and Somlyo, 1987); mimimum detectable masses (for Fe) of 10^{-19} g (thermionic gun) and 10^{-20} g (field-emission gun) and resolution in X-ray scanning images of 9 nm (field-emission gun) (Somlyo and Shuman, 1982).

Although I am concerned primarily in this paper with a history preceding the modern era, I should at least mention some of the important recent developments.

With respect to quantitative analysis, there have been many noteworthy advances since 1970. For thick specimens, analysts now often take P_x/B_x as the basic signal for an element x, where P_x is the characteristic X-ray count and B_x is the background under the characteristic peak (Cobet, 1972; Statham and Pawley, 1978; Boekestein *et al.*, 1984). This procedure gives considerable compensation for the irregular effects of self-absorption in bulk specimens with uneven surfaces.

For the quantitative analysis of thin specimens, at least five developments must be noted:

1. The "peripheral standard" method has been used very successfully by the Munich physiology group for the measurement of electrolyte concentrations, mmol of element per unit volume, in freeze-dried cryosections (Dörge *et al.*, 1975). With this method, local concentrations are determined by comparison with the X-ray signals from standards which are contiguous with the specimen and sectioned simultaneously. The method was also applied earlier by Pallaghy (1973) to the analysis of resin-embedded sections.

2. Several schemes have been used to measure local mmol of element per unit water in soft tissues (for review, see Hall and Gupta, 1986).

3. A formalism has been set up to determine mmol of element/kg wet mass from measurements on freeze-dried cryosections (Zglinicki and Bimmler, 1987).

4. As an alternative to mass normalisation by means of the X-ray continuum count, procedures have been developed for mass normalisation based on various electron signals: attenuation of the transmitted beam (Halloran *et al.*, 1978; Ilyin and Pozsgai, 1979; Linders *et al.*, 1982); attenuation of the zero-loss peak in the electron energy-loss spectrum (Leapman *et al.*, 1984); and intensity of backscattered electrons (Linders and Hagemann, 1983).

5. One can now obtain fully quantitative digital X-ray images, as is discussed extensively elsewhere in this volume.

I have not tried to deal with the unhappy subject of beam damage. It was established by Bahr *et al.*, as long ago as 1965, that irradiation of (uncooled) thin organic films usually induces a mass-loss of 20-30% after an exposure of only ≈ 0.5 10^{-10} Coulomb/μm^2. In favourable circumstances, cooling to about 77 K can reduce mass-loss sensitivity by factors up to 100 (Egerton, 1980 a,b). But from the existing data, one is generally not able to predict the degree of mass loss occurring during actual biological analyses, especially when the specimen is cooled. Hence, in practical quantitative analysis one generally assumes that standards and specimen have been similarly affected, or else one assumes or shows , with cooled specimens, that analyses are completed before much loss has occurred. Beyond this makeshift strategy, progress in understanding and controlling mass loss is difficult because it is difficult to establish the temperature of the specimen itself during irradiation, and because sensitivity may depend on conditions in the column environment (the fate of ejected secondary electrons for example).

I have also not mentioned electron energy-loss spectrometry (EELS) as an alternative to the detection of characteristic X rays. Microanalysis by EELS was proposed, examined in detail theoretically, and tested in a preliminary way forty-five years ago (Hillier, 1943; Hillier and Baker, 1944). It was then dormant for more than twenty years, and taken up again vigorously around 1966 (see Wittry *et al.*, 1969). At its present stage of development, as Shuman and Somlyo (1987) have shown, EELS is far superior to EPXMA in terms of minimum detectable elemental amounts and concentrations, so that EELS should be the method of choice for the identification of tiny localisations. But as these authors point out, quantitation is difficult since elements in their usual biological concentrations appear in EELS spectra as small "blips" on top of a complicated background. For biology, EPXMA is a functioning tool; EELS now stands perhaps where EPXMA was in the 1960's.

References

Andersen CA (1967a) An introduction to the electron probe microanalyzer and its application to biology. In: Glick D (ed) Methods of biochemical analysis, vol 15. Interscience, New York, p. 147

Andersen CA (1967b) The quality of x-ray microanalysis in the ultra-soft x-ray region. Br J Appl Physics 18:1033-1043

Appleton TC (1972) "Dry" ultrathin frozen sections for electron microscopy and X-ray microanalysis: the cryostat approach. Micron 3:101-105

Bahr GF, Johnson FB, Zeitler E (1965) The elementary composition of organic objects after electron irradiation. Lab Invest 14:1115-1133

Boekestein A, Thiel F, Stols ALH, Bouw E, Stadhouders AM (1984) Surface roughness and the use of a peak to background ratio in the X-ray microanalysis of bio-organic bulk specimens. J Microsc 134:327-333

Brooks EJ, Tousimis AJ, Birks, LS (1962) The distribution of calcium in the epiphyseal cartilage of the rat tibia measured with the electron probe X-ray microanalyzer. J Ultrastruct Res 7:56-60

Cobet U (1972) Quantitative electron beam analysis of biological thick tissue samples. In: Third Internat Conf on Medical Physics, Göteborg, Sweden, p. 32.4
Dörge A, Rick R, Gehring K, Mason J, Thurau K (1975) Preparation and applicability of freeze-dried sections in the microprobe analysis of biological soft tissue. J Microsc Biol Cell 22:205-214
Duncumb P (1959) The x-ray scanning microanalyzer. Br J Appl Physics 10:420-427
Duncumb P (1966) Precipitation studies with EMMA- a combined electron microscope and X-ray microanalyser. In: McKinley TD et al. (eds) The electron microprobe. John Wiley & Sons, New York, p 490
Echlin P, Taylor SE (1986) The preparation and X-ray microanalysis of bulk frozen hydrated vacuolate plant tissue. J Microsc 141:329-348
Egerton RF (1980a) Chemical measurements of radiation damage in organic samples at and below room temperature. Ultramicroscopy 5:521-523
Egerton RF (1980b) Measurement of radiation damage by electron energy-loss spectroscopy. J Microsc 118:389-399
Forrest QG, Marshall AT (1976) Comparative X-ray microanalysis of frozen-hydrated and freeze-substituted specimens. In Ben-Shaul Y (ed) Electron microscopy 1976, vol II. Tal International, Jerusalem, p 218
Gupta BL (1976) Water movement in cells and tissues. In: Davies PS (ed) Perspectives in experimental biology, vol 1. Pergamon Press, Oxford, p 25
Gupta BL, Hall TA (1981) The x-ray microanalysis of frozen-hydrated sections in scanning electron microscopy: an evaluation. Tissue & Cell 13:623-643
Gupta BL, Hall TA (1982) Electron probe X-ray microanalysis. Baker PF (ed) Techniques in cellular physiology, pub P128. Elsevier, Amsterdam, p 1
Gupta BL, Berridge MJ, Hall TA, Moreton RB (1978) Electron microprobe and ion-selective micro-electrode studies of fluid secretion in the salivary glands of *Calliphora*. J Exp Biol 72:261-284
Hall TA (1961) X-ray fluorescence analysis in biology. Science 134:449-455
Hall TA (1968) Some aspects of the microprobe analysis of biological specimens. In: Heinrich KFJ (ed) Quantitative electron microprobe analysis. National Bureau of Standards (USA), Washington DC, p 269
Hall TA (1971) The microprobe assay of chemical elements. In: Oster G (ed) Physical techniques in biological research, 2nd edn, vol 1A. Academic Press, New York, p 157
Hall TA (1979) Biological X-ray microanalysis. J Microsc 117:145-163
Hall TA, Gupta BL (1986) EDS quantitation and application to biology. In: Joy DC et al. (eds) Principles of analytical electron microscopy. Plenum Press, New York, p 219
Hall TA, Werba P (1971) Quantitative microprobe analysis of thin specimens: continuum method. In: Proc 25th Anniversary Meeting of EMAG. Institute of Physics, Bristol, p 146
Hall TA, Hale AJ, Switsur VR (1966) Some applications of microprobe analysis in biology and medicine. In: McKinley TD et al. (eds) The electron microprobe. John Wiley & Sons, New York, p 805
Hall TA, Röckert HOE, Saunders RLdeCH (1972) X-ray microscopy in clinical and experimental medicine. Charles C. Thomas, Springfield.
Halloran BP, Kirk RG, Spurr AR (1978) Quantitative electron probe microanalysis of biological thin sections: the use of STEM for measurement of local mass thickness. Ultramicroscopy 3:175-184
Hillier J (1943) On microanalysis by electrons. Physical Rev 64:318-319
Hillier J, Baker RF (1944) Microanalysis by means of electrons. J Appl Physics 15:663-675
Höhling HJ, Hall TA (1969) Elektronenstrahlmikroanalyse als quantitative histologische Methode. Die Naturwissenschaften 56:622-629
Höhling HJ, Hall TA, Boothroyd B, Cooke CJ, Duncumb P, Fitton-Jackson S (1967)

Untersuchungen der Vorstadien der Knockenbildung mit Hilfe der normalen und electronenmikroskopischen electron probe X-ray microanalysis. Die Wissenschaften 54:142-143

Ilyin NP, Pozsgai I (1979) Quantitative microanalysis of thin samples in the EMMA (electron microscope + microanalyxzer). Mikrochimica Acta Suppl 8:213-228

Ingram MJ, Hogben, AM (1968) Procedures for the study of biological soft tissue with the electron microprobe. In: Developments in applied spectroscopy, vol 6. Plenum Press, New York, p 43

Jones RT, Johnson RT, Gupta, BL, Hall TA (1979) The quantitative measurement of electrolyte elements in nuclei of maturing erythrocytes of chick embryo using electron-probe X-ray microanalysis. J Cell Sci 35:67-85

Leapman RD, Fiori CE, Swyt CR (1984) Mass thickness determination by electron energy loss for quantitative X-ray microanalysis in biology. J. Microsc. 133:239-253

Lechene CP (1974) Electron probe microanalysis of picoliter liquid samples. In: Hall TA et al. (eds) Microprobe analysis as applied to cells and tissues. Academic Press, London, p 351

Lechene CP, Bonventre JV, Warner RR (1979) Electron probe analysis of frozen-hydrated bulk tissues. In: Lechene CP & Warner RR (eds) Microbeam analysis in biology. Academic Press, New York, p 409

Lever JD, Duncumb P (1961) The detection of intracellular iron in rat duodenal epithelium. In: Boyd JD et al (eds) Electron microscopy in anatomy. Edward Arnold, London, p 278

Linders PWJ, Hagemann P (1983) Mass determination of thin biological specimens using backscattered electrons. Ultramicroscopy 11:13-20

Linders PWJ, Stols ALH, van de Vorstenbosch RA, Stadhouders AM (1982) Mass determination of thin biological specimens for use in quantitative electron probe X-ray microanalysis. Scanning Electron Microscopy/1982 (IV):1603-1615

Marshall AT (1982) Application of $\phi(\rho Z)$ curves and a windowless detector to the quantitative X-ray microanalysis of frozen-hydrated bulk biological specimens. Scanning Electron Microscopy /1982 (I):243-260

Marshall DJ, Hall TA (1966) A method for the microanalysis of thin films. In: Castaing R et al. (eds) Optique des rayons X et microanalyse. Hermann, Paris, p 374

Marshall DJ, Hall TA (1968) Electron-probe X-ray microanalysis of thin films. Br J Appl Physics 1:1651-1656

Pallaghy CK (1973) Electron probe microanalysis of potassium and chloride in freeze-substituted leaf sections of *Zea mays*. Austral J Biol Sci 26:1015-1034

Robertson AJ, Rivers D, Nagelschmidt G, Duncumb P (1961) Stannosis. Benign pneumoconiosis due to tin dioxide. Lancet, 20 May:1089-1093

Roinel N, de Rouffignac Ch (1982) X-ray analysis of biological fluids: contribution of microdroplet technique to biology. Scanning Electron Microscopy/1982 (III):1155-1171

Russ JC, McNatt E (1969) Copper localization in cirrhotic rat liver by scanning electron microscopy. In: Arceneaux CJ (ed) 27th Proc Electron Microscopy Society Amer. Claitor's, Baton Rouge, p 38

Shuman H, Somlyo AP (1987) Electron energy loss analysis of near-trace-element concentrations of calcium. Ultramicroscopy 21:23-32

Somlyo AP, Shuman H (1982) Electron probe and electron energy loss analysis in biology. Ultramicroscopy 8:219-234

Somlyo AP, Somlyo AV, Devine, CE, Peters PD, Hall TA (1974) Electron microscopy and electron probe analysis of mitochondrial cation accumulation in smooth muscle. J Cell Biol 61:723-742

Somlyo AP, Bond M, Somlyo AV (1985) Calcium content of mitochondria and endoplasmic reticulum in liver frozen rapidly *in vivo*. Nature (London) 314:622-625

Sims RT, Hall TA (1968) X-ray emission microanalysis of proteins and sulphur in rat plantar epidermis. J Cell Sci 3:563-572
Statham P, Pawley JB (1978) A new method for particle X-ray microanalysis based on peak to background measurements. Scanning Electron Microscopy/1978 (I):469-478
Swift JA (1979) Minimum depth electron probe X-ray microanalysis as a means for determining the sulphur content of the human hair surface. SCANNING 2:83-88
Tousimis AJ, Adler I (1963) Electron probe X-ray microanalyzer study of copper within Descemet's membrane in Wilson's disease. J Histochem Cytochem 11:40-47
Wittry DB (1969) Recent advances in instrumentation for microprobe analysis. In: Möllenstedt G & Gaukler KH (eds) X-ray optics and microanalysis. Springer, Heidelberg, p 206
Wittry DB, Ferrier RP, Cosslett VE (1969) Microanalysis in the transmission electron microscope by selected area electron spectrometry. In: Möllenstedt G, Gaukler KH (eds) X-ray optics and microanalysis. Springer, Heidelberg, p 293
Wroblewski J, Wroblewski R (1986) Why low tempereature embedding for X-ray microanalytical investigations? A comparison of recently used preparation methods. J Microsc 142:351-362
Zglinicki T von, Bimmler M (1987) The intracellular distribution of ions and water in rat liver and heart muscle. J Microsc 146:77-85
Zs.-Nagy I, Lustyik G, Zs.-Nagy V, Zarandi B, Bertoni-Freddari C (1981) Intracellular Na:K ratios in human cancer cells as revealed by energy dispersive X-ray microanalysis. J Cell Biol 90:769-777
Zs.-Nagy I, Lustyik G, Bertoni-Freddari C (1982) Intracellular water and dry mass content as measured in bulk specimens by energy-dispersive X-ray microanalysis. Tissue Cell 14:47-60

Conference Discussion

B.L. Gupta: When talking about beam damage in cryosections, do you not think that different components are differentially sensitive to mass loss damage according to their chemical composition? E.g. polysaccharides are more sensitive than proteins and extracellular substance more than cell compartments?

T.A. Hall: Yes, very likely. In analysis at moderate resolution, you can monitor the continuum and stop analyzing before much loss of mass occurs. In analysis at high resolution you have to pass that point and I suppose some systematic errors to occur because of variable sensitivity.

K. Zierold: Could you make some suggestions on how to reduce beam damage during microanalysis, perhaps by coating sections with special materials?

T.A. Hall: The data on the effectiveness of cold-stages are very inconsistent. In my own lab stage-cooling seems to have been quite useful, while coating with Al or C has not been (except when very high probe currents are used and the specimen gets hot, but this does not happen any more).

B.L. Gupta: In the last illustration of the liver tissue (Russ and McNatt, 1969, In: Arcenaux, C.J. (ed.) 27th Proc. EMSA (USA), Claitor's, Baton Rouge, p. 38) it is interesting to note that the highest peak for Ca is in endoplasmic reticulum. It has taken 20 years to realize that endoplasmic reticulum is a major Ca store.

J. Shelburne: Could you please review the earliest electron microprobe studies of human disease states? (e.g. Wilson's Disease, etc.)

T.A. Hall: I think these are references to early studies in the appendix of a 1971 paper: Hall, T.A. (1971) In: Oster G (ed) Physical Techniques in Biological Research, 2nd Edition, Vol. 1A, Academic Press, New York, p. 157.

(See also Hall, T.A. et al., 1972) X-ray microscopy in clinical and experimental medicine. Charles C. Thomas (pub.), Springfield, Illinois).

P. Ottensmeyer: You expressed the feeling that present analytical techniques for electron probe X-ray microanalysis have sufficient spatial resolution and sensitivity. Are there then no questions in the field of microanalysis of cellular physiological processes that would be aided by, or even require, the development of analytical techniques with improved spatial resolution and sensitivity?
T.A. Hall: Undoubtedly, there are many problems where better resolution and sensitivity would be valuable. For most of these problems, beam-induced alteration of structure must be an obstacle at some level of resolution. Perhaps, the study of the early stages of mineralization is an especially favourable topic, since the essential structures may be largely inorganic and relatively less prone to disarray. Brij Gupta has pointed out that when you are dealing with ordered structures, composite images like yours might be obtainable with lower dose through the use of optical filtering or equivalent computer processing. (And my impression is that cross correlation techniques can be similarly useful even for non-periodic objects.)

Specimen Preparation and Other Limitations in Quantitative Eletron Probe X-Ray Microanalysis (EPXMA) Using Ultrathin Sections

Norbert Roos
Electron Microscopical Unit for Biological Sciences
University of Oslo, P.O.Box 1062, N-0316 Oslo 3, Norway

INTRODUCTION

Qualitative and quantitative EPXMA of biological soft tissue quite often requires preparation methods that prevent or reduce loss or dislocation of elements. Conventional specimen preparation methods for transmission electron microscopical studies involve penetration of the specimen with fluids for fixation, dehydration, embedding in plastic and polymerisation at high temperatures and sectioning. Such preparation routines entail a risk of redistribution and loss of elements (Harvey et al., 1976; Coetzee and Van der Merwe, 1984), which will vary from one specimen to another but sometimes not interfere with the logic of the study at all. Conventional specimen preparation methods have been used to characterize, for example, asbestos fibres in lung tisue (Champness et al., 1976) or the accumulation of heavy metals in kidney tissue (Galle et al., 1968).

However, in physiology almost all X-ray microanalytical studies deal with highly diffusible elements, which by definition are prone to displacement. In addition one is usually concerned with the problem of measuring the concentration of a given element rather than establishing its presence or absence i.e., special preparative techniques are required if meaningful, fully-quantitative, analysis is to be achieved. Even in the presence of precipitating agents, liquid fixation is not adequate for physiological X-ray microanalysis (Simson et al., 1979).The methods of choice are the "low temperature", or "cryo"-preparations, where small pieces of tissue, cell suspensions, or monolayers of cultured cells, are frozen very rapidly (quench-frozen) and processed further. If the specimen is to be sectioned different options are available:

<u>Plastic embedding of frozen-dried or frozen-substituted specimens</u>

The water in the bulk frozen specimen is either removed by freeze-drying and replaced by a resin (Ingram and Ingram, 1980), or replaced by a cold organic solvent, which is then in turn displaced by an infiltrating resin (Harvey, 1982). After resin polymerization the blocks can be cut at ambient temperatures on a dry knife. New resin formulations permit the infiltration of dehydrated frozen tissue blocks at temperatures in the range 200K to 250K ; polymerization of these methacrylate resins can be performed by U.V. at for example 250K (for reviews see Robards and Sleytr, 1985; Steinbrecht and Müller, 1987). The localization of solutes in matrix-free domains such as extracellular spaces and vacuoles is precluded in frozen-dried preparations. The translocation of solutes during the resin infiltration of frozen-dried and frozen-substituted specimens is a serious potential artefact.

Cryosectioning of frozen-hydrated specimens

Frozen-hydrated blocks of soft biological tissue can be sectioned easily on commercially available equipment. Sections can be frozen-dried before being transferred to the microscope (Somlyo and Silcox, 1979) or they can be transferred to the microscope in the fully-hydrated state using a cold transfer device, and frozen-dried in the column (Zierold, 1982). Analysis of frozen-dried sections is employed in the majority of quantitative studies in physiology (Rick et al., 1982; Roomans and Wei, 1985), and it is commonly accepted that the use of a cold transfer device to keep the temperature low during freeze-drying (i.e. "internal" as opposed to "external" freeze-drying), reduces the risk of lipid melting and redistribution of elements (Hagler and Buja, 1986) and radiation damage during analysis (Zglinicki and Uhrík, 1988).

Alternatively the sections can be analysed in the fully-hydrated state, a procedure currently subject to severe limitations. Imaging frozen-hydrated specimens is very difficult, since contrast is poor (unlike conventionally prepared specimens, no electron-dense stains are present because their application and presence is incompatible with the requirements of analysis). An additional disadvantage is that fully-hydrated specimens are extremely beam sensitive (Dubochet et al., 1988). The electron dose needed to obtain adequate counting rates for X-ray microanalysis will thus lead to the destruction of the specimen during analysis. Another limitation of the method is the poor performance of the standard cold traps in electron microscopes. They are designed to reduce contamination of the specimen kept at ambient temperatures. If the specimen is kept at very low temperatures (< 140K), the temperature difference between specimen and cold trap is too small to prevent contamination.

MATERIALS AND METHODS

Perfusion of rat pancreas

Male rats (200gr), deprived of nourishment overnight, were anaesthetized with Nembutal, dissected and a cannula inserted in the aorta. The pancreas was perfused using a pump delivering 20ml/minute of perfusate to obtain steady-state conditions. The perfusate was a Krebs-Ringer phosphate buffer (148 mmol/l Na^+, 7 mmol/l K^+, 1 mmol/l Ca^{2+}, 1.4 mmol/l Mg^{2+}, 118 mmol/l Cl^-, 22 mmol/l $HPO_4^{2-}/H_2PO_4^-$, 1.4 mmol/l SO_4^{2-}) containing 4% dextran to substitute for plasma proteins and additives (e.g. carbachol) or deletions (e.g. Ca^{2+}) at 310K. The pancreas was either frozen in situ or carefully removed intact except for blood and nerve supplies and mounted on a frame under perfusion medium in order to be frozen with the helium-slammer.

Cryofixation

Rat exocrine pancreas was either mounted on a frame under perfusion medium, excess medium removed and quench-frozen against a helium-vapor cooled copper block (Escaig, 1984) or quench-frozen in situ between two polished liquid nitrogen-cooled copper blocks attached to pliers (Hagler and Buja, 1984).

The samples were immediately transferred to liquid nitrogen for long-term storage.

Freeze-drying, fixation and plastic-embedding

Small fragments of the frozen tissue were transferred to the specimen holder of a freeze-drying unit designed and constructed in our laboratory following the general requirements defined by Ingram and Ingram (1984). The unit consists of a glass cylinder (condenser) containing the specimen holder, the temperature of which can be monitored and regulated. The glass cylinder is placed in liquid nitrogen. A vacuum of better than 10^{-5} Torr was provided by the vacuum system of an old Siemens Elmiskop IA. The pressure was measured at the gauge situated in the vacuum system of the microscope. The freeze-drying followed mainly a schedule proposed by Ingram and Ingram (1984).

Aldehyde-fixation was achieved by heating para-formaldehyde powder to 333K (3 hours) in the condenser. Osmium fixation was performed by introducing osmium-tetroxide crystals (i.e. a valve sealing off a small glass balloon containing the crystals was opened).

In order to introduce the embedding medium (Spurr low-viscosity resin or EPON) the vacuum was broken by leaking dry nitrogen gas into the unit. The resin was allowed to infiltrate for 3 hours and the blocks cured at 333K overnight.

Freeze-substitution and plastic-embedding

Small vials with dry acetone (dried over a molecular sieve) containing 2% osmium-tetroxide were frozen in liquid nitrogen. Frozen tissue fragments were transferred onto the frozen acetone and the vials then surrounded by dry ice in a thermally insulated box. The dry ice was allowed to sublime overnight, which led to a slow warm-up of the sample. The dehydrated specimen were then embedded in Spurr low-viscosity resin and cured overnight.

Sectioning of resin embedded specimens

Sections for X-ray microanalysis were cut using a dry knife on a Sorvall MT 5000 ultramicrotome. The sections had a nominal thickness of 200-400 nm. The sections were picked-up with an eyelash attached to a straw and transferred to a 75 or 100 mesh copper grid previously coated with a Formvar/carbon film. A second thin (ca. 10 nm) carbon film was evaporated onto the grid containing the sections.

Sections for ultrastructural examinations were cut onto water and post-stained with uranyl acetate and lead citrate.

Cryosectioning and cold transfer

Cryosections were cut on a Sorvall MT 5000 and FS 1000 cryo-attachment.

Nominal thickness was 100-250 nm and the temperature for specimen and knife was set to between 128K and 138K. Sections were mounted on Formvar/carbon coated copper grids (50, 74 and 100 mesh) using an eyelash. The grids were then moved to the grid holder, sandwiched under a second Formvar film and the grid holder transferred to a pre-cooled Gatan cold stage. The cold stage was inserted into the column and the cryosections frozen-dried at 193K for 20 minutes.

Analysis and quantitation

Analysis was done in the scanning transmission mode using a JEOL 100 CX electron microscope equipped with a free condenser lens 1 control and a LINK 860 series 2 energy-dispersive X-ray spectrometer (resolution 148 eV). The detector was mounted horizontally and the sample therefore tilted 35^0 with respect to the detector. The nominal distance between detector and specimen was 24 mm. Acceleration voltage was 80 kV and the zero strobe peak and copper K line were routinely used to calibrate the spectrometer for gain and zero shifts.

Spectra were collected over the range of 0-20 keV for 50 or 100s livetime at a resolution of 20 eV/channel. The output count rate was adjusted to approximately 1000 cps excluding the zero strobe peak. The selected area raster mode was routinely used at a scan speed of 1 or 10s per frame.

Quantitation was performed using the QUANTEM/FLS software provided by LINK systems, based on the continuum method developed by Hall and collaborators (Hall, 1979; Hall and Gupta, 1983).

Aminoplastic standards were used as cryo-standards (Roos and Morgan, 1985).

Each experimental group consisted of 3-4 animals. A maximum of 8 cells per animal were chosen, and three different spectra from each organelle-type (i.e. mitochondria, zymogen granules, endoplasmic reticulum and apical cytoplasm), were taken in each cell and averaged i.e. a total of approximately 72 to 96 spectra for each organelle in each treatment group were taken.

RESULTS AND DISCUSSION

Cryofixation

Cryoprocedures provide the only means of obtaining adequate structural information, to instantly arrest all physiological activity, and to stabilize the specimen. All cryoprocedures have the first crucial step in the preparative chain in common: cryofixation or quench-freezing. It must guarantee a high cooling rate (> 4000K/s), should be relatively simple and reproducible and should yield a frozen specimen that is easily processed further. Most freezing methods are reviewed in an article by Plattner and Bachmann (1982).

First attempts to freeze excised and trimmed pieces of rat exocrine pancreas by mechanically plunging them into liquid propane resulted in relatively well frozen tissue (unpublished results). However, in order to obtain well frozen specimen, the volume of the specimen should be kept below 1 mm^3. This requirement jeopardized our attempts to freeze the tissue in a well defined physiological state, since excision and excessive trimming could possibly introduce artefacts (sudden fall of blood pressure and drying out of the small tissue cubes respectively).

We therefore resorted to freezing our specimen by slamming it against a helium-vapor cooled highly-polished copper block (helium-slammer: described by Escaig, 1984)). This method allowed us to excise the entire pancreas whilst perfusing it, mount it onto a frame and quench freeze the central part of it in less than 2 minutes. The freezing quality was excellent, reproducible, and ultrastructural details were preserved far into the tissue (Fig. 1), even though no cryoprotectant was used. All the frozen samples were routinely checked for the quality of the freezing procedure by freeze-substitution as described in the Materials and Methods section. At a later stage a freezing method was adopted that allowed freezing of the pancreas in situ . Two small copper blocks, attached to a pair of pliers, were cooled in liquid nitrogen and the pancreas squeezed between them without prior dissection. Freezing with the pliers was not as reproducible as with the helium slammer, since the pressure applied to the specimen would vary each time, but yielded a specimen with two well frozen surfaces and a central region with some ice-crystal damage. In most cases the dimension of the ice-crystal damage was smaller than the spatial resolution of the analysis.

Freeze-drying and freeze-substitution

Freeze-drying of tissue fragments has been performed in a "home made" freeze-drying unit especially designed to permit control of the specimen temperature, fixation of the dehydrated sample by para-formaldehyde and/or osmic acid vapor, resin embedding and to avoid possible rehydration (breaking the vacuum of the unit with dry nitrogen gas). Morphological preservation was very good (Fig. 2) and all the intracellular compartments such as nuclei, mitochondria, rough endoplasmic reticulum and zymogen granules were easily identified. Dehydration and resin infiltration was obviously complete, since no empty spaces in the tissue or the individual cells could be seen and the cell and organelle boundaries were distinct and not blurred.

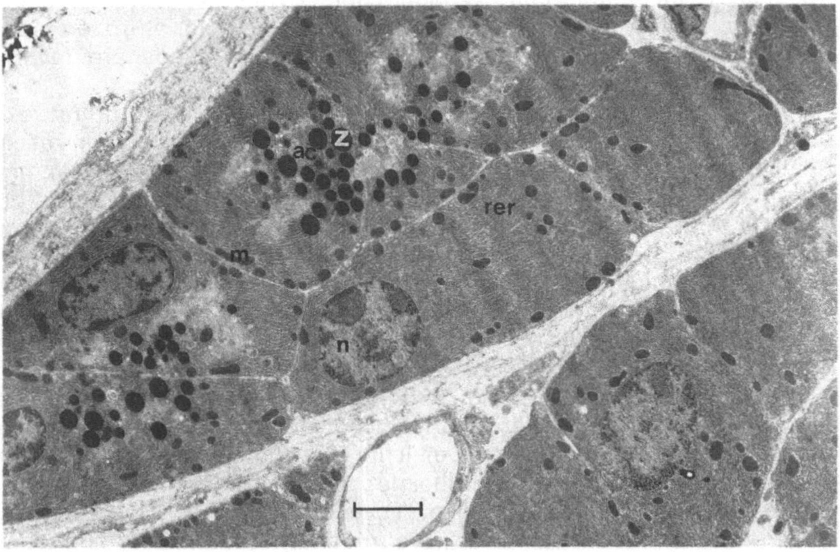

Figure 1: Transmission micrograph of an ultrathin section of rat exocrine pancreas, quench-frozen and freeze-substituted. Section post-stained with uranyl acetate/lead citrate. z = zymogen granules, ac = apical cytoplasm, n = nucleus, rer = rough endoplasmic reticulum, m = mitochondria. Bar = 3 μm.

Freeze-drying has to be performed at temperatures above the recrystallization temperature for frozen water (140K) i.e. any vitrified region of the specimen will be devitrified and ice-crystals will start growing during the process of freeze-drying. However, we did not find ice-crystal damage in cells close to the "freezing-front",but increasingly more damage towards the center of the sample where freezing was presumably slower i.e. the original ice-crystal damage pattern was preserved, indicating that the process of freeze-drying did not contribute considerably to the ice-crystal damage, unless one assumes a "healing" effect of the resin.

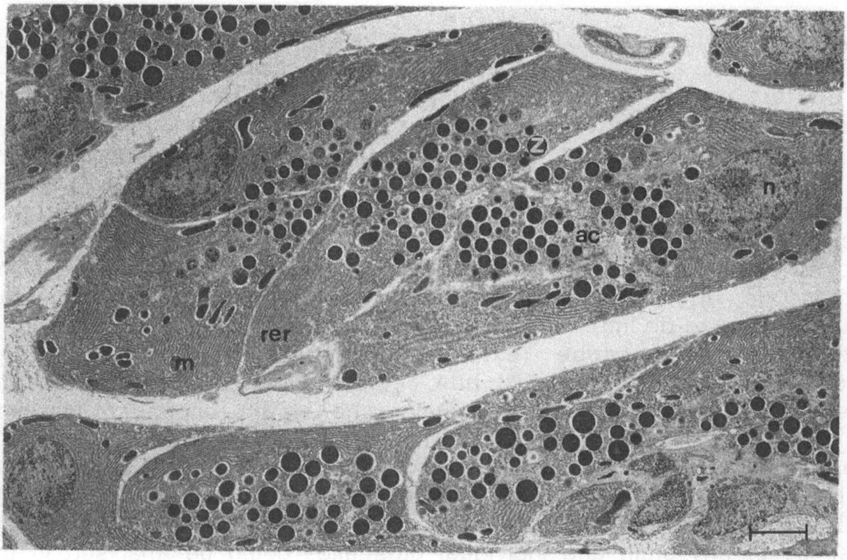

Figure 2: Transmission micrograph of an ultrathin section of quench-frozen, freeze-dried, formaldehyde-vapour-fixed, plastic-embedded rat exocrine pancreas. Section post-stained with uranyl acetate/lead citrate. z = zymogen granules, ac = apical cytoplasm, n = nucleus, rer = rough endoplasmic reticulum, m = mitochondria. Bar = 3 µm.

Although freeze-drying is a well established method for morphological applications it has seldom been used for analysis of ions even though Ingram and Ingram (1980) submit evidence suggesting that intracellular ion compartments can be preserved.

Freeze-substitution of quench-frozen tissue includes a slow replacement of ice by anhydrous organic solvents such as acrolein/ether mixtures or acetone at low temperatures. Again this process will take place above the devitrification temperature of water. However, it has been shown that ice-crystal growth during freeze-substitution is relatively slow (Steinbrecht, 1985) and, for the substitution times employed in our studies, negligible. Ice-crystal damage patterns in freeze-substituted samples are virtually the same as in freeze-dried samples. This means that no, or extremely little damage occurs at the surface that made contact with the cold metal mirror and that increasing damage occurs towards the center of the sample. In the well frozen region organelles could be identified easily (Fig. 1).

Cryoultramicrotomy and cryotransfer

Cryosections are very susceptible to compression and chatter as well as the formation of crevasses (Chang et al., 1983). Holes and cracks are also all too easily introduced during the handling of the sections with an eyelash (Fig. 3) Badly frozen tissue (i.e. tissue containing excessive ice-crystals) hardly sections at all. Ultrathin fully-hydrated cryosections display poor amplitude contrast in the electron microscope and are easily damaged by the beam. To increase contrast and to stabilise the sections they are freeze-dried in the electron microscope after transfer with a low temperature specimen holder (internal freeze-drying). In some reports freeze-drying arrangements outside the microscope (external freeze-drying) are used (e.g. Wroblewski et al.,1983) but it has been shown recently that external freeze-drying is inadequate, especially when the sections are warmed to room temperature. Damage is manifested as the melting of lipid droplets (Hagler and Buja, 1986) with the resultant redistribution of highly mobile electrolytes (Roos and Barnard, unpublished results).

Contrast in freeze-dried cryosections (Fig. 3) was very good in the absence of any stain and individual organelles could be analysed in the scanning transmission mode. For the purpose of subcellular quantitative electron-probe X-ray microanalysis, the morphology of the cryosections does not impose serious constraints other than those introduced by the freezing process itself.

Analytical results

In order to obtain reasonably reliable analytical data that reflect the elemental composition present in vivo we had to prepare the specimen in different ways, and to measure and quantitatively compare identical subcellular compartments. The latter objective turned out to be a relatively complex task, since the quantitative results obtained are expressed in mmol/kg plastic for the resin embedded samples and mmol/kg dry weight of biological tissue for the freeze-dried cryosections. An additional problem with resin embedded samples is that the degree of penetration of the resin into different compartments of the cell is not known.

In both cases it was not possible to deduct physiological mmol/l values since we did not know the degree of hydration for the different subcellular compartments.

It could clearly be demonstrated that freeze-substituted samples lost potassium from all the subcellular compartments measured, except the zymogen granules, compared to freeze-dried samples. The very low phosphorus values for rough endoplasmic reticulum in freeze-substituted compared to freeze-dried samples suggest a loss or redistribution of this element during freeze-substitution with acetone. The freeze-substitution schedule employed does not seem to preserve the original elemental composition faithfully. Longer (lasting up to 3 weeks) and more stringent schedules (Marshall, 1980; Harvey, 1980), yet no quantitative data on the subcellular level, have been reported. Retention of the elements might improve by using more sophisticated substitution protocols. However, substitution procedures that last for 3 weeks were not considered practical given the number of replicates required to meet statistical demands in most biological investigations.

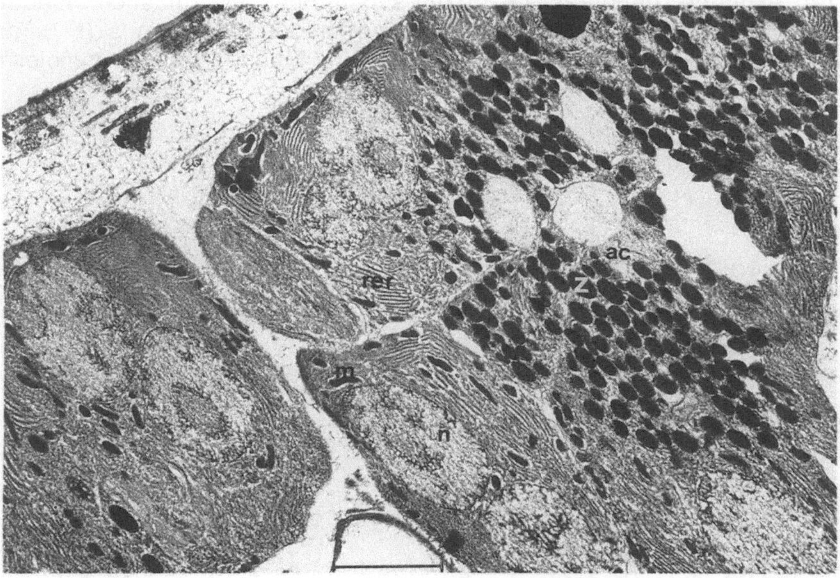

Figure 3. Transmission micrograph of a thin frozen-dried cryosection of quench-frozen rat exocrine pancreas. z = zymogen granules, ac = apical cytoplasm, n = nucleus, rer = rough endoplasmic reticulum, m = mitochondria. Bar = 3 µm.

Elemental loss is less pronounced in freeze-dried resin-embedded samples. However it is still considerable compared to freeze-dried cryosections, especially the loss of phosphorus and potassium from the apical and basal cytoplasm (Table 1). One possible explanation is that membrane phospholipids are extracted and redistributed by the highly hydrophobic Spurr and Epon resins rendering the exposed membranes "leaky" for potassium and other highly mobile elements.

Elemental gradients between different subcellular compartments were most pronounced in freeze-dried cryosections. Furthermore the potassium:sodium ratios were also very pronounced in all the organelles in these preparations.The lower potassium:sodium ratios in resin-embedded material seems to be mainly a result of potassium loss during specimen preparation. Standard deviations of the measurements in plastic for FD and FS specimens and mmoles/kg dry wt. for CS specimens. ZG = zymogen granules, AC = apical cytoplasm, ER = basal cytoplasm, M = mitochondria.

Table 1: Elemental composition of different subcellular compartments in rat exocrine pancreas in frozen-substituted and plastic-embedded specimens (FS), frozen-dried and plastic-embedded specimens (FD), and frozen-dried cryosections (CS) (Roos, 1988). Mean ± SD, in mmoles/kg

		COMPARTMENT			
		ZG	AC	ER	M
FS	Na	32 ± 21	27 ± 24	36 ± 21	31 ± 19
	Mg	8 ± 8	14 ± 11	12 ± 9	3 ± 4
	P	36 ± 9	68 ± 28	150 ± 17	85 ± 24
	S	80 ± 23	41 ± 18	31 ± 10	42 ± 13
	Cl*	-	-	-	-
	K	29 ± 9	23 ± 6	27 ± 9	24 ± 6
	Ca	5 ± 3	1 ± 1	2 ± 1	2 ± 1
FD	Na	62 ± 27	46 ± 20	58 ± 19	38 ± 17
	Mg	20 ± 10	23 ± 11	29 ± 11	15 ± 9
	P	71 ± 30	207 ± 59	308 ± 31	201 ± 53
	S	366 ± 76	99 ± 19	58 ± 6	93 ± 13
	Cl*	-	-	-	-
	K	112 ± 65	120 ± 21	167 ± 21	132 ± 29
	Ca	25 ± 6	4 ± 2	3 ± 2	2 ± 2
CS	Na	105 ± 9	344 ± 29	358 ± 18	200 ± 8
	Mg	39 ± 2	157 ± 14	217 ± 16	102 ± 9
	P	126 ± 14	1844 ± 126	2760 ± 147	1188 ± 29
	S	674 ± 21	383 ± 20	215 ± 15	404 ± 19
	Cl	96 ± 8	318 ± 28	357 ± 23	156 ± 6
	K	121 ± 6	1161 ± 98	1554 ± 72	801 ± 24
	Ca	54 ± 2	37 ± 1	15 ± 3	14 ± 2

* Cl not measured since also present in embedding medium.
FS: n = 6 cells; FD: n = 27 cells; CS: n = 24 cells

cryosections were also consistently lower (unpublished observations). The results unequivocally indicate that freeze-dried cryosections of the tissue of interest yield the more reliable and reproducible data.

We therefore concluded, that freeze-dried cryosections offer the best preparations to obtain quantitative data on the elemental composition of subcellular compartments.

REFERENCES

Champness PE, Cliff G, Lorimer GW (1976) The identification of asbestos. J Microsc 108: 231-249

Chang J-J, MacDowell AW, Lepault J, Freemann R, Walter CA, Dubochet J (1983) Freezing, sectioning and observation artefacts of frozen-hydrated sections for electron microscopy. J Microsc 132: 109-123

Coetzee J, Van der Merwe CF (1984) Extraction of substances during glutaraldehyde fixation of plant cells. J Microsc 135: 147-158

Dubochet J, Adrian M, Chang J-J, Homo J-C, Lepault J, McDowell A, Schulz P (1988) Cryo-electron microscopy of vitrified specimens. Q Rev Biophys 21: 129-228

Escaig J (1984) Control of different parameters for optimal freezing conditions. In: Revel JP, Barnard T, Haggis GH (eds) The science of biological specimen preparation. Scanning Electron Microsc Inc., AMF O'Hare, Illinois, pp 117-122

Galle P, Berry JP, Stuve J (1968) New applications of electron microprobe analysis in renal pathology. Proc. 3rd Nat. Conf. Electron Microprobe Analysis, Electron Probe Analysis Society of America, Chicago, Paper 41

Hagler HK, Buja LM (1984) New techniques for the preparation of thin freeze-dried cryosections for X-ray microanalysis. In: Revel J-P, Barnard T, Haggis GH (eds) The science of biological specimen preparation. Scanning Electron Microsc Inc. AMF O'Hare, Illinois, pp 161-166

Hagler HK, Buja LM (1986) Effect of specimen preparation and section transfer techniques on the preservation of ultrastructure, lipids and elements in cryosections. J Microsc 141: 311-317

Hall TA (1979) Biological X-ray microanalysis. J Microsc 117: 145-163

Hall TA, Gupta BL (1983) The localization and assay of chemical elements by microprobe methods. Q Rev Biophys 16: 279-339

Harvey DMR (1980) The preparation of botanical samples for ion localization studies at the subcellular level. Scanning Electron Microsc II: 409-419

Harvey DMR (1982) Freeze-substitution. J Microsc 127: 209-221

Harvey DMR, Flowers TJ, Hall JL (1976) Localization of chloride in leaf cells of the halophyte Suaeda maritima by silver precipitation. New Phytol 77: 319-323
Ingram FD, Ingram MJ (1980) Freeze-dried, plastic-embedded tissue preparation: a review. Scanning Electron Microsc IV: 147-160
Ingram FD, Ingram MJ (1984) Influences of freeze-drying and plastic embedding on electrolyte distributions. In: Revel JP, Barnard T. Haggis GH (eds) The science of biological specimen preparation. Scanning Electron Microsc Inc. AMF O'Hare, Illinois, pp 167-174
Marshall AT (1980) Sections of freeze-substituted specimens. In: Hayat MA (ed) X-ray microanalysis in Biology. McMillan Publishers Ltd, London, pp 207-239
Plattner H, Bachmann L (1982) Cryofixation: a tool in biological ultrastructural research. Int Rev Cytol 79: 237-304
Rick R, Dörge A, Thurau K (1982) Quantitative analysis of electrolytes in frozen-dried sections. J Microsc 125: 239-247
Robards AW, Sleytr UB (1985) Low temperature methods in biological electron microscopy. In: Glauert AM (ed) Practical methods in electron microscopy, vol 10. Elsevier, Amsterdam New York Oxford
Roos N (1988) A possible site of calcium regulation in rat exocrine pancreas cells: An X-ray microanalytical study. Scanning Microsc 2: 323-329
Roos N, Barnard T (1985) A comparison of subcellular element concentrations in frozen-dried, plastic-embedded, dry-cut sections and frozen-dried cryosections. Ultramicroscopy 17: 335-344
Roos N, Morgan AJ (1985) Aminoplastic standards for electron-probe X-ray microanalysis (EPXMA) of ultrathin, frozen-dried cryosections. J Microsc 140: RP3-RP4
Roomans GM, Wei X (1985) X-ray microanalysis of resting and stimulated rat pancreas. Acta Physiol Scand 124: 353-359
Simson JAV, Bank HL, Spicer SS (1979) X-ray microanalysis of pyroantimonate-precipitable cations. Scanning Electron Microsc II: 779-792
Somlyo AV, Silcox J (1979) Cryoultramicrotomy for electron probe analysis. In: Lechene C, Warner R (eds) Microbeam analysis in biology. Academic Press, New York, pp 535-550
Steinbrecht RA (1985) Recrystallization and ice-crystal growth in a biological specimen, as shown by a simple freeze substitution method. J Microsc 140: 41-46
Steinbrecht RA, Müller M (1987) Freeze-substitution and freeze-drying. In: Steinbrecht RA, Zierold K (eds) Cryotechniques in biological electron microscopy. Springer, Berlin Heidelberg New York London Paris Tokyo, pp 149-172

Wroblewski J, Müller RM, Wroblewski R, Roomans GM (1983) Quantitative X-ray microanalysis of semi-thick cryosections. Histochemistry 77: 447-463

Zglinicki T, Uhrík B (1988) X-ray microanalysis with continuous specimen cooling: is it necessary? J Microsc 151: 43-47

Zierold K (1982) Cryopreparation of mammalian tissue for X-ray microanalysis in STEM. J Microsc 125: 149-156

Conference Discussion

L. Edelmann: Did you try different freeze-substitution and freeze-drying procedures? Do you see trends for improved ion retention using other embedding techniques?
N. Roos: We tried different freeze-drying schedules (with freeze-drying times up to 7 days) and vapour-fixations (formaldehyde- and/or osmium-vapour fixation) (Roos and Barnard, 1985) but only one freeze-substitution procedure. The retention of mobile elements could possibly be improved, but fully-quantitative data comparing cryosectioned with plastic-embedded specimens using more stringent freeze-substitution schedules and/or low temperature embedding media have not been published yet.

A. Warley: Do you think that the differences that you see in quantitation between resin-embedded specimens and cryosections could relate to differences in beam behaviour between cryosections, sections of resin embedded samples, and the aminoplastic standards?
N. Roos: We have reported earlier (Roos and Morgan, 1985) that aminoplastic standards may be used for determining element concentrations in a range of plastic-embedded specimens and in cryosectioned tissue samples, since they are equally stable under electron irradiation as the "cryostandards" using gelatine or dextran organic matrices.

B.L. Gupta: When comparing the validity of results from cryosections dried externally or internally one must ask the question, "What are the results valid against"? Do you agree?
N. Roos: I do. In both cases one is left with an unique set of data. However, one is tempted to believe that the more sophisticated piece of equipment (a cold transfer device) might give the more reliable data!

B.L. Gupta: When comparing different types of preparations for quantitation should one not include Z^2/A corrections? These may be substantially different for frozen-dried v/s embedded and fixed v/s unfixed specimens?

N. Roos: The corrections are included in the QUANTEM/FLS software used in this study.

Th. von Zglinicki: The comparison between freeze-substituted, freeze-dried embedded and freeze-dried sections would become more convincing if you measured the local water content and give the results from freeze-dried sections as concentrations per wet mass.

R.A. Steinbrecht: Assuming that in plastic-embedded, freeze-dried specimen all tissue water is replaced by plastic it would be easy to calculate wet weight concentrations and compare them with dry weight concentrations obtained from freeze-dried cryosections. Couldn't we estimate the possible deviations from this assumption in practice and still use this method to calculate wet weight concentrations and dry mass content?

R. Rick: I would like to make a comment. If we want to be heard by people outside of our field, we should make every effort to provide them with numbers they can relate to, i.e. concentrations per liter cell water. I realize that none of the methods available for estimating the water content is perfect, but nevertheless we should do it.

N. Roos: I do agree that wet weight concentrations might appeal to people outside our field. However, estimation of wet weight concentrations from plastic-embedded specimens is difficult, since we do not know to what degree the resin penetrates the different subcellular compartments. The problem could possibly be solved by adding a known concentration of say Br to the resin and by measuring its distribution within the cell, but we still would not know whether the water is entirely replaced by the resin or not. To estimate the degree of hydration of the different subcellular compartments in frozen-dried cryosections can be done in different ways. Peripheral standards are a possible solution, but are not applicable where tissue has to be frozen rapidly in situ as in this study. Another possibility is to measure the optical density of the frozen-hydrated section and compare it with the frozen-dried section. This method cannot cope with the differential shrinkage of the subcellular compartments during the freeze-drying process.

Since we did not perform such measurements on our specimens, we cannot make any statements concerning the degree of hydration of the subcellular compartments in rat exocrine pancreas.

Freeze-Substitution and Low Temperature Embedding for Analytical Electron Microscopy

L. Edelmann
Fachr. 3.5 Med. Biologie
Universität des Saarlandes
D-6650 Homburg-Saar FRG

Introduction

It is generally accepted that rapid freezing of biological material yields preparations in which the cellular water, the cellular ultrastructure and the cellular ions are captured in a life-like state. Due to the fact that electron microscopic analysis of morphology and ion distribution of frozen-hydrated preparations still poses severe technical problems (Zierold, 1986) different techniques are commonly used to obtain stable dehydrated specimens which can be analysed routinely in an electron microscope. Among these follow-up procedures freeze-substitution (FS) followed by embedding in a plastic yields specimens which are particularly well suited for morphological investigations (Steinbrecht and Müller, 1987). Although it has been claimed that meaningful elemental microanalysis can also be carried out with sections of freeze-substituted and embedded biological material (for references see Zierold and Steinbrecht, 1987) it is generally assumed that freeze-dried cryosections are best suited for microanalysis and that the retention of mobile ions at or near their original sites appears very unlikely after FS and embedding. Even if a considerable amount of mobile ions can be detected in membrane enclosed cells or subcellular compartments substantial ion redistribution artefacts cannot be ruled out.

The only way to prove that FS and embedding of biological material can be carried out without causing ion redistribution is to demonstrate that the ions under investigation are localized at identical cellular sites in the rapidly frozen and in the freeze-substituted and embedded preparations. The present paper describes a method for ion detection which is applicable to frozen-hydrated cryosections and sections of plastic embedded material. Experimental evidence is provided for the view that properly freeze-substituted and low temperature embedded material can be used for electron microscopic analysis of both morphology and subcellular ion distribution.

Theoretical Considerations

It is well known that most transmission electron micrographs of ultrathin sections of embedded biological material represent the distribution of heavy metal ion deposits at specific cellular structures. Electron-dense particles usually accumulate during chemical fixation (e.g. with OsO_4) or after sectioning of the plastic embedded material during the staining procedure (e.g. with uranyl acetate and lead citrate). However, accumulation of electron-dense ions in living cells is also possible. Many physiologists have studied the accumulation of Rb^+, Cs^+, and Tl^+ in living cells with the aim to better understand the molecular mechanism of K^+ ion accumulation in living cells as well as the molecular mechanisms responsible for electrophysiological phenomena observed on living cells (for review see Edelmann, 1984a). These studies have unequivocally shown that Rb^+, Cs^+, and Tl^+ all accumulate in living cells according to the same mechanism as K^+ which is the main cellular cation of most living cells. In particular, experiments with living isolated frog muscle revealed that more than 80 % of the cellular K^+ can be replaced reversibly with the heavy Cs^+ or Tl^+ ions. This means that we can obtain living cells in which the main cellular cation is Cs^+ or Tl^+ instead of K^+. Similar to the behaviour of K^+ in normal cells the intracellular Cs^+ or Tl^+ ions are in constant exchange with extracellular Cs^+ or Tl^+ ions and the dying cell loses the accumulated Cs^+ or Tl^+ ions during chemical fixation. However, the living Cs^+- or Tl^+-loaded cell can be cryofixed and we thus achieve a preparation which is particularly well suited for electron microscopic investigations as it contains a large amount of electron-dense particles. Provided these particles are unevenly distributed in the cell either at specific subcellular structures or within membrane enclosed compartments we should be able to visualize directly this uneven distribution in frozen-hydrated cryosections by transmission electron microscopy.

Experiments of this kind have already been carried out (Edelmann, 1984b, 1988a) and results are shown in Fig. 1. When comparing the frozen-hydrated cryosections of Tl^+-loaded muscle (Fig. 1a, 1c) with those of normal K^+-containing muscle (Fig. 1b) it was found that the electron-dense Tl^+ ions are mainly localized in the A bands and at Z lines; more precisely, the Tl^+ ions are accumulated at individual protein filaments of the muscle preparation (Fig. 1c). The intense protein staining caused by Tl^+ implies that most of the cellular Tl^+ ions are bound to the proteins and not dissolved in the surrounding water; otherwise one would expect a poor or even a negative staining of the proteins. These results confirm findings which had previously been obtained with freeze-dried and embedded muscles and also with freeze-dried cryosections (for review see Edelmann, 1984a). The results are seen as a confirmation of the association-induction hypothesis (Ling, 1984) according to which most of the cellular cations are weakly bound to β- and γ-carboxyl groups of cellular proteins.

Now we may ask whether cation distributions can be found in freeze-substituted and plastic embedded muscle which are similar to or identical with those shown on frozen-hydrated preparations. Considering the findings that most of cellular cations are lost after FS and embedding at high temperatures (Roos and Barnard, 1986) and that preparational artefacts caused by organic solvents are reduced when the specimen is kept at low temperatures during dehydration and embedding (Kellenberger, 1986) one may speculate that cellular cations may eventually be preserved at their protein binding sites if FS, embedding, and polymerization of the biological material are all carried out at the lowest possible temperatures. In the past mainly

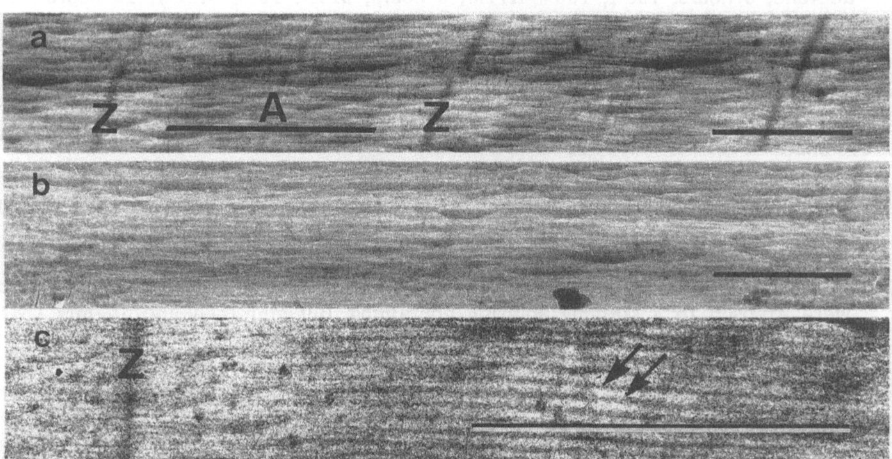

Fig 1a - c. Frozen-hydrated cryosections of frog sartorius muscles. **a** and **c** Tl^+-loaded muscles. **b** normal K^+-containing muscle. Preferential sites of Tl^+-accumulation are found at myosin filaments (arrows) in the A bands (A) and at Z lines (Z). Bars: 1 µm

diethylether was used as substitution medium when the cellular retention of mobile ions was required. But since FS with diethylether is extremely slow at 193K (Steinbrecht and Müller, 1987) most of the frozen cellular water was usually substituted at much higher temperatures with the result of ice crystal growth and hence structural artefacts. In order to avoid such artefacts the FS to be described in the next section was carried out with acetone solely at 193K and Lowicryl K11M was chosen to render feasible embedding of the substituted material below 213K.

Specimen Preparation and Results

Frog sartorius and semitendinosus muscles were isolated from Northern American leopard frogs (Rana pipiens pipiens, Schreber) and used either when containing a normal amount of K^+, or after the major part of cell K^+ (about 80%) was replaced by Cs^+ or Tl^+ (Ling, 1977). Resting or contracting muscles were cryofixed by rapid contact with a LN_2 cooled copper block as described elsewhere (Edelmann, 1988b). Small pieces of frozen tissue were transferred into the REICHERT-JUNG CS auto cryosubstitution unit. FS was carried out in pure acetone for 8 days at 193K. Afterwards the muscles were infiltrated with Lowicryl K11M (70% K11M + 30% acetone, 5 hours, 193K; 100% K11M, 5 hours, 213K; 100% K11M, 12 hours, 213K), and transferred into fresh embedding medium in flat embedding moulds. After sealing the moulds tightly with UV translucent covers the specimens were polymerized by UV irradiation at 213K for 24 hours. All these steps were carried out inside the CS auto. The polymerized specimens were warmed up to room temperature (without further UV irradiation) and removed from the moulds. DIATOME diamond knives were used for obtaining ultrathin and 0.1 µm thick sections, glass knives for 0.3 µm thick wet-cut and dry-cut sections. Ultrathin sections were stained with uranyl acetate and lead citrate. 0.1 µm thick sections were exposed to a staining solution containing 100 mM LiCl and 10 mM CsCl as described elsewhere (Edelmann, 1984, p. 884).

Different properties of the freeze-substituted muscle preparations are shown in the following figures: Fig. 2 demonstrates the quality of structure preservation of resting and contracting muscle after FS for 8 days in pure acetone at 193K followed by low temperature embedding (LTE). Fig. 3 shows that thin sections of similarly prepared muscle can be stained with the electron-dense Cs^+ ion. Fig. 4 reveals distribution patterns of Cs^+ and Tl^+ in differently prepared muscles (see figure legends); here the electron-dense ions have accumulated in the living muscles before cryofixation.

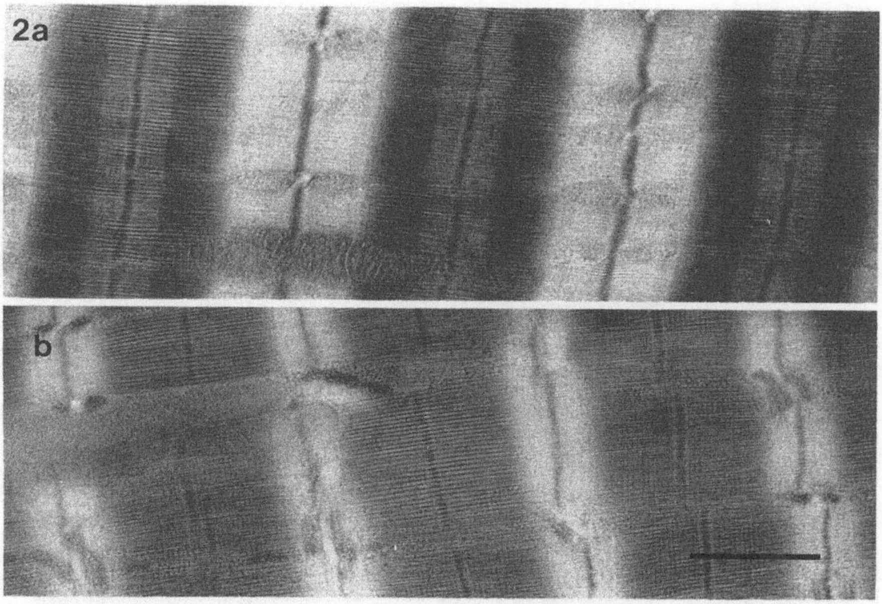

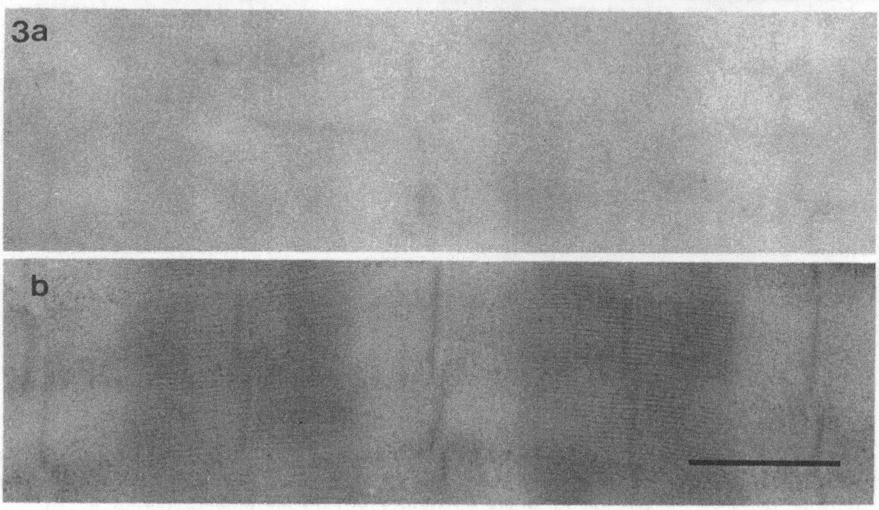

Fig 2a, b. Ultrathin sections of Cs$^+$-loaded frog semitendinosus muscles after FS and LTE. **a** Resting muscle. **b** Contracting muscle (isotonic contraction), cryofixed after 1 sec of electrical stimulation at a rate of 50/sec. Bar: 1 µm

Fig 3a, b. 0.1 µm thick sections of frog sartorius muscles after FS and LTE, **a** without any staining, **b** after staining with Cs$^+$ (see text). Z lines and myosin filaments are stained by Cs$^+$. Bar: 1 µm

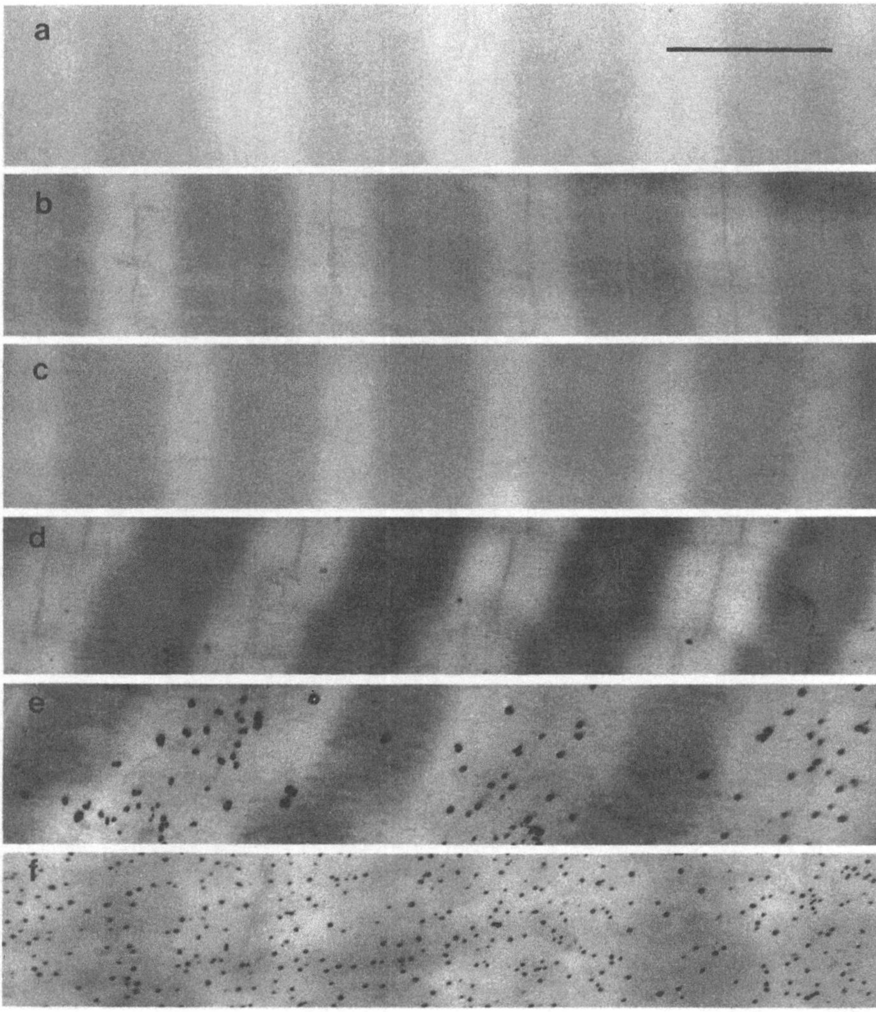

Fig 4a - f. 0.3 µm thick sections of frog semitendinosus muscles after FS and LTE. **a** Wet-cut section of a Cs^+-loaded muscle. **b** Dry-cut section of a Cs^+-loaded muscle. **c** Dry-cut section of a Cs^+-loaded contracting muscle (same as shown in Fig. 2b). **d** Dry-cut section of a Tl^+-loaded muscle. **e, f** Dry-cut sections of electrically stimulated Tl^+-loaded muscles; these muscles were stretched and kept at constant length during stimulation. The poor contrast in **a** indicates loss of electron-dense Cs^+-ions during wet-cutting. **b** and **d** show preferential accumulation of electron-dense Cs^+ and Tl^+ ions at A bands and Z lines of resting muscle. Note the poor contrast in **c** (see text) and the differently distributed Tl-precipitates in **e** and **f** (see text). Bar: 2 µm

Discussion

The results presented here show the following: Firstly, biological material can be stabilized by FS and LTE at temperatures below 213K in such a way that the subcellular ultrastructure is maintained. It must be mentioned however, that when acetone is used as substitution medium it requires rather a long time to complete the FS process. When muscles were substituted for only 2 - 4 days at 193K and subsequently infiltrated in K11M it was impossible to obtain well polymerized specimens. Apparently, a considerable amount of cellular water can only be substituted slowly at 193K and at least 1 week is necessary to dehydrate the muscle sufficiently such as to render feasible the successful infiltration with K11M and subsequent polymerization.

Secondly, once we have obtained a well preserved muscle which has only been exposed to acetone and K11M at temperatures below 213K the embedded cellular proteins reveal a remarkable property: they are able to bind alkali-metal ions (Fig. 3). Electron-dense Cs^+ ions are accumulated at the very protein sites (preferentially myosin and Z line proteins) which bind the main cellular cation in the living cell (Fig. 1). Since chemically fixed and conventionally prepared muscles cannot be stained with electron-dense alkali-metal ions (Edelmann, 1984a) we may conclude that the physico-chemical properties of the proteins responsible for cellular cation-binding are not drastically altered during FS and embedding at low temperatures. The extent of modification of these properties caused by different solvents at varying low temperatures remains to be determined in future studies.

In the third place, the results show that mobile cellular ions like Cs^+ and Tl^+ are retained in the resting muscle at the same subcellular sites (A bands and Z lines) after FS and LTE where they are found in frozen-hydrated preparations (compare Figs. 4b, d with Fig. 1a). Two different Tl-distributions are shown in Figs. 4e and 4f. The micrographs were obtained from Tl^+-loaded muscles which had been cryofixed during contraction. The formation of the visible Tl-containing precipitates is interpreted as follows: During contraction Tl^+ ions are liberated from their binding sites; they diffuse and combine with free anions (e.g. phosphate ions). Depending on the different contraction-states of the muscle fibers at the moment of cryofixation the precipitates are differently distributed within the muscle cells. Liberation and redistribution of the main cellular cation during muscle contraction are postulated in a new model for muscle contraction derived from principles of the association-induction hypothesis (Ling, 1984). Further evidence for the redistribution of cellular cations during contraction is provided by Fig. 4c: The contracted Cs^+-loaded muscle shows a very poor contrast between A and I bands and the Z lines are not stained with the electron-dense Cs^+ ions. An interpretation of this finding is that liberated Cs^+ ions do not precipitate but diffuse away from their original sites either before cryofixation or during FS and/or LTE.

The results discussed so far leave several problems for future investigations. Although the uneven distribution of electron-dense cations in frozen-hydrated and in freeze-substituted preparations of the resting muscle is alike we do not yet know whether all cellular cations can be retained during FS and LTE. Quantitative X-ray microanalysis is required to determine the exact amount of cellular cations retained after FS and LTE (some preliminary experiments with normal K^+-containing muscles yielded high K-peaks similar to those obtained from freeze-dried and K11M-embedded preparations, see Edelmann, 1986). The comparison of microanalytical data gained from resting and contracting muscle both after cryosectioning and FS and LTE would be of particular interest. Such comparative studies may help to answer the following questions: 1) Is it possible to confirm the results concerning the observed redistribution of cellular cations during muscle contraction by independent methods? 2) What is the composition of the precipitates seen in Figs. 4e, f? 3) Are the precipitates formed before cryofixation or during FS and LTE? 4) Is it possible to determine free and bound cellular cations by examination of freeze-substituted and embedded preparations? From a theoretical point of view such a determination is conceivable: It is very unlikely that ions freely dissolved in cellular water can be preserved at their original sites during the replacement of the cellular frozen water with organic solvents. Either the ions precipitate and may eventually be found at cellular mechanical barriers or the ions are soluble in the organic solvent and one may expect their diffusion out of the open muscle cells (broken after cryofixation) during FS and/or LTE. On the other hand, the distribution patterns of Cs^+ and Tl^+ seen in frozen-hydrated and in freeze-substituted resting muscles strongly support the view that most of the weakly bound ions can be retained at their original binding sites.

Conclusion

Freeze-substitution of biological material in acetone at 193K and low temperature embedding in Lowicryl K11M at 213K yields stable preparations well suited for sectioning and subsequent morphological and microanalytical studies. It is expected that FS and LTE can be optimized such as to render possible the quantitative evaluation of free and bound mobile ions in living cells.

References

Edelmann L (1984a) Subcellular distribution of potassium in striated muscles. Scanning Electron Microsc 1984/II : 875 - 888

Edelmann L (1984b) Frozen-hydrated cryosections of thallium-loaded muscle reveal subcellular potassium binding sites. Physiol Chem Phys and Med NMR 16 : 499 - 501

Edelmann L (1986) Freeze-dried embedded specimens for biological microanalysis. Scanning Electron Microsc 1986/IV : 1337 - 1356

Edelmann L (1988a) The cell water problem posed by electron microscopic studies of ion binding in muscle. Scanning Microscopy 2 : 851 - 865

Edelmann L (1988b) The contracting muscle: a challenge for freeze-substitution and low temperature embedding. In: Albrecht RM, Ornberg RL (eds) The science of biological specimen preparation, 1988. SMI, AMF O'Hare, IL 60666, in press

Kellenberger E, Carlemalm E, Villiger W (1986) Physics of the preparation and observation of specimens that involve cryoprocedures. In: Müller M, Becker R, Boyde A, Wolosewick J (eds) The science of biological specimen preparation, 1985. SEM, AMF O'Hare, IL 60666, pp 1 - 20

Ling GN (1962) A physical theory of the living state. The association-induction hypothesis. Blaisdell, Waltham MA

Ling GN (1977) Thallium and cesium in the muscle cells compete for the adsorption sites normally occupied by K^+. Physiol Chem Phys 9 : 217 - 222

Ling GN (1984) In search of the physical basis of life. Plenum Publishing Corp, New York

Roos N, Barnard T (1986) Preparation methods for quantitative electron probe X-ray microanalysis of rat exocrine pancreas: a review. Scanning Electron Microsc 1986/II : 703 - 711

Steinbrecht RA, Müller M (1987) Freeze-substitution and freeze-drying. In: Steinbrecht RA, Zierold K (eds) Cryotechniques in biological electron microscopy. Springer, Berlin Heidelberg New York, pp 149 - 172

Zierold K (1986) Preparation of cryosections for biological microanalysis. In: Müller M, Becker R, Boyde A, Wolosewick J (eds) The science of biological specimen preparation, 1985. SEM, AMF O'Hare, IL 60666, pp 119 - 127

Zierold K, Steinbrecht RA (1987) Cryofixation of diffusible elements in cells and tissues for electron probe microanalysis. In: Steinbrecht RA, Zierold K (eds) Cryotechniques in biological electron microscopy. Springer, Berlin Heidelberg New York, pp 272 - 282

Acknowledgements. This work was supported by a grant from the city of Homburg-Saar. I would like to thank Mrs. B. Reiland for typing the manuscript.

Conference Discussion

B.L. Gupta: I am very pleased to note that other people are also now finding grossly differential (non-uniform) distribution of monovalent ions in cells in situ. We reported this with low resolution in epithelia (Gupta and Hall, 1979, Fed. Proc. 38:144-153). The ion selectivity of fixed negative charges such as ß- and γ-COO^- critically depends on their stereo-chemical positions in native state. Can one assume that Tl^+ is "binding" to the same sites as K^+ in I and A bands?

A. Dörge: Would you follow from the uneven cellular distribution you found for Cs^+ and Tl^+ that also K^+ is unevenly distributed under control conditions?

M.F. Wendt-Gallitelli: Do you have quantitative data supporting your suggestion that the inhomogeneities you find in the sarcomeres in Cs^+ or Tl^+ distribution in freeze-dried and freeze-substituted muscle exactly correspond to inhomogeneities in K^+ distribution in cryosections?

L. Edelmann: From the following experimental findings it is concluded that Tl^+, Cs^+ and K^+ are adsorbed (weakly bound) with different adsorption energies to the same negative fixed charges of the muscle cell: Tl^+, Cs^+, and K^+ can stoichiometrically and reversibly displace each other in living frog muscle cells. The amounts of cations (Tl^+, Cs^+, or K^+) accumulated in the cells at given extracellular concentrations are quantitatively described by adsorption isotherms. The selectivity in cation accumulation ($Tl^+ > K^+ > Cs^+$) is similar in intact cells and in effectively membrane-less muscle cell preparations, and the cation accumulation does not follow predictions of the Donnan membrane theory (Ling, 1977). Experimental evidence for the conclusion that the different ions accumulate at the same sites was provided by X-ray microanalysis in freeze-dried cryosections of Tl^+-, Cs^+-, Rb^+-loaded and of normal K^+-containing muscle (Edelmann, 1983, Physiol. Chem. Phys. 15:337-344). It was qualitatively shown that the different cations are all preferentially localized in the A bands and at Z lines. These findings are qualitatively similar to the uneven distributions of electron-dense ions seen in freeze-dried or freeze-substituted and embedded or frozen-hydrated muscle preparations. Recently, quantitative X-ray microanalysis on freeze-dried cryosections of heart muscle cells confirmed the finding of a preferential (about 2-fold) K^+ accumlation in the A band compared to the I band (Zglinicki, 1988, Gen. Physiol. Biophys. 7:495-504). Whether the uneven distributions of electron-dense ions in plastic embedded preparations correspond exactly to the uneven K^+ distribution in freeze-dried or frozen-hydrated cryosections remains to be determined.

Th. von Zglinicki: Your results demonstrate nicely the distribution of negative fixed charges in muscle. However, ions will be precipitated to the fixed charges during drying. Could you summarize the evidence you have that they cannot travel over distances larger than, say, the length of the I band during drying? Are the results obtained from freeze-dried sections on an ultrastructural level meaningful?

L. Edelmann: It is well known that K^+ ions migrate faster on the surface of glass that bears anionic sites than in dilute solution (reviewed by Ling, 1984, p. 243). Hence, if we assume that K^+ accumulation at cellular proteins is due to a "mobile adsorption" at anionic sites (comparable to K^+ adsorption at glass surfaces) the conclusion is justified that the adsorption energies between charges of opposite sign may be influenced by the freeze-drying procedure such as to facilitate or to impede the redistribution of K^+. In order to rule out the possibility of the uneven distributions of K^+, Rb^+, Cs^+, and Tl^+ found in freeze-dried preparations being freeze-drying artefacts the distribution of cellular cations was investigated with frozen-hydrated preparations. The Tl^+ distribution found in frozen-hydrated cryosections and the Rb^+ and Cs^+ distributions found in quick-frozen single muscle fibers by autoradiography of ^{86}Rb and ^{134}Cs at liquid nitrogen temperature (Edelmann, 1980, Histochemistry 67:233-242) are consistent with those results obtained from freeze-dried cryosections which showed a preferential accumulation of K^+ in the A bands and at Z lines. The fact that so far very few laboratories have confirmed a preferential accumulation of K^+ in the A band of freeze-dried muscle-cryosections may point to the problem of freeze-drying in such a way that redistribution of mobile ions is minimized during freezing-drying or during electron-irradiation of the dried cryosections.

A. Dörge: Can you comment on the height of the K gradients you would expect to exist within a cell?
L. Edelmann: If cellular K^+ accumulation is due to a weak binding (mobile adsorption) to β- and γ-carboxyl groups of cellular proteins the local K^+ concentrations follow the local concentrations of fixed anionic sites in the cell. The striated muscle is particularly well suited to test the concept of ion adsorption because about 2/3 of the β- and γ-carboxyl groups of the muscle cell are concentrated at myosin in the A band. Hence, in stretched muscle one would expect at least a 2-fold higher K^+ concentration in A bands than in I bands. Much higher concentration gradients are expected between e.g. a myosin filament and the surrounding water. Although such K^+ gradients cannot yet be measured, a quantitative determination of local Tl^+ concentrations on an ultrastructural level in Tl^+-loaded cells appears feasible by densitometry of micrographs obtained from frozen-hydrated cryosections (Fig. 1c).

R. Rick: Those cationic binding sites seem to accept a variety of different monovalent cations. What is the selectivity of these sites for K^+ as compared to Na^+?
L. Edelmann: The association-induction hypothesis describes quantitatively the intracellular Na^+ and K^+ concentrations at varying external ion concentrations as sums of ions dissolved in cellular water and ions adsorbed at protein sites (Ling, 1984, chap. 11). According to the adsorption isotherms used in these descriptions the K^+/Na^+ selectivity is in the order of 100 in frog muscle. The selective uptake of different cations by proteins in sections of freeze-dried and embedded muscle preparations has been studied with the LAMMA technique and a K^+/Na^+ selectivity of over 10 was found (Edelmann, 1980, Physiol. Chem. Phys. 12:509-514). It remains to be investigated whether it is possible to freeze-dry or freeze-

substitute and embed the tissue in such a way that the embedded proteins accumulate K^+ with the high selectivity postulated by the association-induction hypothesis.

H.K. Hagler: With reference to the precipitates shown in Fig. 4e and f, what are the statistics of this observation, i.e. how many preparations were studied and how often was the precipitate found to be as illustrated.
L. Edelmann: More than 20 electrically stimulated muscle cells have been inspected and Tl precipitates were always visible. In most cases these precipitates were distributed as shown in Fig. 4e. Occasionally the distribution pattern shown in Fig. 4f was seen. A small percentage of the resting Tl^+-loaded muscles investigated so far revealed also precipitates in the I band. Whether this is due to an unintended stimulation of muscle cells during preparation for cryofixation remains to be determined.

H.K. Hagler: What type or types of ice were present in the fully hydrated sections shown in Fig. 1? Was this measured using electron diffraction?
L. Edelmann: So far, perfect diffuse electron diffraction patterns corresponding to vitreous ice were found only in extracellular areas of the fully hydrated sections. Electron diffraction of the well frozen parts of the muscle areas usually yielded very faint rings (indicating cubic ice) in a predominantly diffuse diffraction pattern. This finding may be due to the fact that frog sartorius muscle is always covered by a few micrometers of extracellular water (connective tissue) and that, therefore, the cooling during cryofixation is not fast enough to yield perfect vitreous ice.

H.K. Hagler: It is now well accepted that hydrated sections followed by freeze-drying offer the best preservation of mobile ions. The technique of cryosectioning is much quicker and has been shown to give reproducible results. What compelling reason necessitate the development of alternate plastic techniques which severely limit the X-ray sensitivity in terms of peak/

background ratio?

L. Edelmann: Reproducible results are not necessarily free of artefacts. Efforts to optimize different competitive preparation methods may ultimately lead to a better understanding of the different artefacts introduced by the different procedures. A few reasons speaking in favour of an optimized embedding tecnique are:

1) A main goal of electron microscopic investigation of biological material is to find out the relations between cellular ultrastructure, properties of cellular macromolecules, and cellular ion distributions. Hence, it appears worthwhile to develop a preparation technique yielding stable preparations which can be used simultaneously for morphological, immunocytochemical, and microanalytical studies.

2) It is very difficult to measure precisely the specimen temperature during freeze-drying of a cryosection. Hence, it is very difficult to establish an optimized generally accepted freeze-drying procedure which can be carried out likewise in different laboratories (see discussion with von Zglinicki). Reproducible dehydration during FS poses no difficulties. Once an optimized FS and embedding procedure is established, permanently available stable preparations can easily be obtained in every laboratory.

3) In many cases sectioning of plastic-embedded material is easier than cryosectioning (e.g. plant material). Particularly, if the preparation includes well frozen and not well frozen areas, it is difficult to obtain uniform cryosections.

4) Several microanalytical techniques require stable sections of even thickness (e.g. EELS, LAMMA, microdensitometry); for these techniques sections of plastic-embedded material are better suited than freeze-dried cryosections.

5) The selective uptake of ions with similar physicochemical properties by cellular proteins can be studied with plastic-embedded material (see discussion with Rick) but not with freeze-dried cryosections.

6) The determination of free and bound cellular ions is eventually feasible with plastic-embedded material (see text) but not with freeze-dried cryosections.

Ensuring the Validity of Results in Biological X-Ray Microanalysis

Thomas von Zglinicki
Institute of Pathology
Charite
Schumannstr. 20/21
GDR - 1040 Berlin

Introduction

The conformation of any macromolecule in a given cell is primarily determined by the interplay of internal and external electric charges. Local water content and ion concentrations are basic regulators of all cellular functions, therefore.

X-ray microanalysis, and especially X-ray microanalysis of frozen dried ultrathin cryosections, is still one of the most promising techniques to measure ion concentrations on an organelle or even suborganelle scale. However, most of the intracellular ions are highly diffusible. Artificial redistribution of ions during the preparation of tissue is highly probable.

In recent years, however, it became possible to preserve the native distribution of ions at least on the organelle level. Critical steps in the preparation together with the experimental evidence for the unbiasedness of results will be reviewed here.

Nativity on a cellular scale

As a very basic requirement, X-ray microanalytical estimations of ion concentrations averaged over the whole cell have to coincide with well- known physiological data. Until recently even this was rather an exception, at least in oxygen-sensitive organs like the heart (Fig. 1).

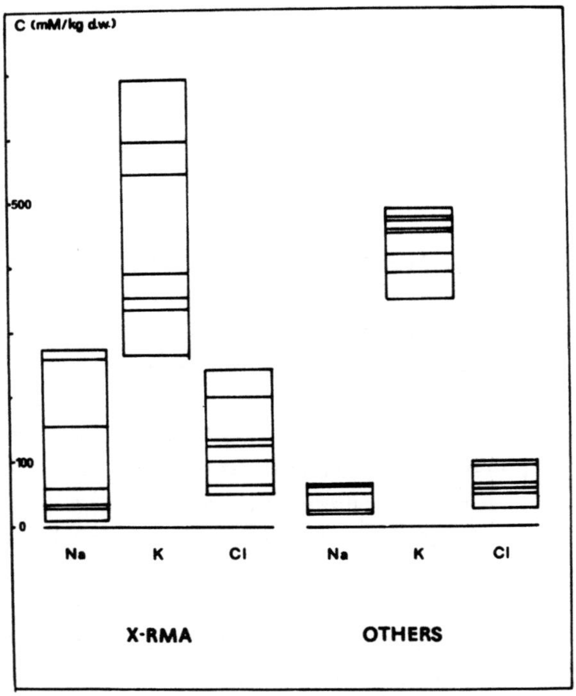

Fig 1. Averaged cellular concentrations of Na, K, and Cl in heart muscle cells as measured by X-ray microanalysis (left side) and by ion- sensitive microelectrodes (K, Cl), radioactive tracer techniques (Cl), Ag-titration (Cl), flame photometry (Na, K), and atomic absorption spectroscopy (Na, K) (right side) in different laboratories. Data are taken from a compilation (von Zglinicki et al. 1986) and updated by incorporation of recent X-ray microanalytical results (Hagler and Buja 1986, von Zglinicki and Bimmler 1987).

The measured concentrations of Na, K, and Cl are in much closer accordance to the expected values if ischemic and traumatic influences during the sampling of the tissue can be minimized (Hagler and Buja 1986, von Zglinicki et al. 1986).

Ionic gradients between intra- and extracellular space will equilibrate quickly by diffusion if the plasma membrane becomes leaky by traumatic influences before freezing.

According to a simple model computation (von Zglinicki et al. 1986) at a typical depth of about 10 um below the surface of a tissue sample ion concentrations will be changed measurably already after less than 10 msec. Therefore, trauma should be avoided at all during tissue sampling or freezing should start simultaneously with the onset of the lesion.

This can be achieved by squeeze-freeze techniques using a modified Wollenberger clamp (Hagler et al. 1983), by cryoballistic methods (Chang et al. 1980, Robards and Sleytr 1985) or by a cryobioptical technique (von Zglinicki et al. 1986). Whilst squeezing methods freeze the tissue from outside, i.e. from the endothelium, both cryoballistic and cryobioptic techniques expose cells within an organ for freezing and subsequent sectioning. The use of cryoballistic projectiles is more reproducible in terms of entrance velocity than that of the hand-held cryobioptic needle. On the other hand, a cryobioptic needle is easily constructed and versatile in use.

Both in the cryoballistic and the cryobioptic technique a chilled hollow needle is propulsed into the tissue, either pressure-powered or by hand. The needle tip has to be as sharp as possible but should not become blunt after repeated use. In our experience, a cryobioptic needle with a tip angle of 15 or more does not give specimen cylinders reproducibly. With 5 - 10 sectioning angle of a needle made from stainless steel one can take about 30 - 50 samples before the needle has to be resharpened.

Even before touching the tissue there is the risk of ischemic injury. An elapsed time of about one minute between killing of the experimental animal and exposure of the organ under study for sampling is quite normal. However, it was shown that ion concentrations are artificially altered if rat liver is removed from the blood supply for only 10 sec (von Zglinicki et al. 1986). The brain, for instance, is known to be even more sensitive against short periods of ischemia, reacting with ion shifts large enough to produce severe extracellular edema (David 1969). Therefore, cervical dislocation is preferred to kill the animals in order to retain at least a residual heart action as long as possible. The organ under study is not removed from the blood vessels before sampling.

Nativity on a subcellular scale

Artificial ion redistributions might occur between intracellular organelles, even if ion exchange between intra- and extracellular space is excluded in the preparation. It is much more complicated to prove whether such artefacts occur: reference values for ion concentrations are in general only available for isolated organelles in media with more or less artificial ion composition. These values might significantly deviate from the ion composition of the same organelles in vivo. Moreover, ions remaining in an organelle after isolation in, e.g., low-ionic media might belong to a less exchangeable fraction. Therefore, the comparison of measurements done on isolated organelles by X-ray microanalysis and independent techniques could be unable to detect artefacts occuring during the preparation of tissue.

On the other hand, the intracellular water distribution is highly sensitive against preparative influences. We decided, therefore, to compare the distribution of water before and after preparation of frozen dried cryosections for X-ray microanalysis (von Zglinicki et al. 1987, von Zglinicki 1988a). Isolated rat liver mitochondria in a solution of bovine serum albumin were choosen as a model system. The water content of the protein solution was adjusted to 74 - 76% similar to that of the cytoplasm. Mitochondria from adult rat livers are known to contain less than 50% water, ensuring a steep water gradient to the medium. The model should be adequate for the intracellular space, therefore.

The water, respective dry mass, content of mitochondria in relation to that of the medium was measured in vitro by interference microscopy (Barer and Dick 1957) and, using frozen dried cryosections, by paired X-ray background counts (Somlyo et al. 1979, Gupta and Hall 1982) and mainly by microdensitometry of electron micrographs (Zeitler 1971, Linders et al. 1981). Segregation structures in the form of bright haloes around mitochondria were found if the freezing rate was low as judged by a mean size of ice crystal remnants in the medium larger than about 100 nm (von Zglinicki 1988a). Those haloes are most probably due to an efflux of subcooled water from mitochondria during slow freezing. None of these haloes and no differences of the relative dry mass of mitochondria before and after preparation for X-ray microanalysis were found using well-frozen samples (von Zglinicki et al. 1987). Local differences in section thickness could be ruled out as a contributing factor by measurements on serial sections.

Two sources of artefacts in X-ray microanalytical estimations of subcellular ion, respective element, distributions are not yet excluded with these results. First, radiation damage in the elctron microscope, especially at ambient temperature, may influence some elements preferentially (Hall and Gupta 1982, Rick et al. 1982, Cantino et al. 1986). Second, ion redistributions without concomittant water shifts might occur, especially during the transfer of frozen dried cryosections through the moist laboratory air into the electron microscope (Hagler and Buja 1986).

To evaluate this, element concentrations in different compartments of rat liver cells were measured on sections cryotransferred into the electron microscope and freeze-dried therein and compared with results obtained on frozen dried cryosections exposed to the room air for 2 min, examined at ambient temperature or at 138 K (von Zglinicki and Uhrik 1988, von Zglinicki et al. 1988). Sulphur losses were found if measured at ambient temperature independent on transfer conditions. This is in agreement with earlier observations on radiation-induced loss of S which can be blocked by cooling the specimen (Cantino et al. 1986).

As a transfer artefact, small-scale ultrastructural changes including loss of image contrast and/or the enlargement of segregation structures were observed in those cryosections which were not carbon-coated before exposure to the air.

Moreover, uncoated sections display significant loss of ultrastructure and gross distorsion of element contents if they are only slightly colder than room temperature during transfer. This in in accordance with recent results obtained on heart muscle (Hagler and Buja 1986). Carbon-coating seems to be sufficient to protect frozen dried cryosections during a short exposure to room air against these detrimental effects. Only minor changes of intracellular ion distributions occur in suitably transferred cryosections (von Zglinicki and Uhrik 1988, von Zglinicki et al. 1988).

The measurement of intracellular water distributions

True ultrastructural analytical resolution is at present not possible using fully hydrated specimens (Zierold 1988). On the other hand, ion concentrations per dry mass as obtained from measurements on frozen dried cryosections are of very limited biological significance. In fact, knowledge of the intracellular water distribution might be a very important piece of information which could help to solve questions concerning, for instance, the state of intracellular water and ions (Cameron et al. 1988, von Zglinicki 1988b) or the mechanisms of cellular aging (von Zglinicki 1987, von Zglinicki 1988a).

There is a number of different techniques in use, basing either on measurements on the same section both in the frozen hydrated and in the frozen dried state (Gupta and Hall 1981, Saubermann and Heymann 1987) or on measurements performed on frozen dried sections only together with some external standard, either incorporated into the same section (Rick et al. 1979) or on a different section of the same thickness (Zierold 1986). Instead of an external water content standard, additional knowledge of the water content of one compartment or of the averaged water content of the whole cell can be used (von Zglinicki and Bimmler 1987).

Each method has its own advantages and disadvantages. Saubermann and Heymann (1987) average the X-ray background over a large area of the frozen hydrated section and measure local P/b values only in the frozen dried state. This minimizes radiation damage as the main problem of analysis of frozen hydrated sections. However, it requires constant section thickness over the whole area and no shrinkage at all of the section during drying.

By measuring both characteristic and background signals also in hydrated cryosections (Gupta and Hall 1981, Gupta and Hall 1982) changes of sections during drying are ruled out. However, one has to be fully aware of the problems of radiation damage and difficult identification of structures in frozen hydrated sections which limit the obtainable resolution to the cellular level.

Measurements on frozen dried cryosections allow better analytical resolution. The local dry mass might be estimated by measuring the X-ray background (Rick et al. 1979, Gupta and Hall 1982), the dark field current (Zierold 1986), or by densitometry of micrographs (Linders et al. 1981, von Zglinicki et al. 1987). For any of these methods, the absence of artificial water shifts as demonstrated above is the basic prerequisite. Shrinkage of specimens during drying and compression of sections do not influence the results as far as these changes are of the same magnitude in all compartments including the dry mass standard, if in use. In fact, the amount of shrinkage might be different in compartments with widely differing water content, e.g. intra- and extracellular space, but the shrinkage seems to be homogeneous in all intracellular compartments in well frozen sections (von Zglinicki et al. 1987, von Zglinicki 1988a).

The relative dry mass estimations obtained can be converted in absolute dry mass or water content values by using external standards. This introduces the requirement of constant section thickness between sample and standard. The incorporation of the standard into the same section, e.g. as an external albumin layer of known water and ion content (Rick et al.

1979) adds the possibility of artificial water and ion fluxes between tissue and standard (Saubermann et al. 1986). These problems can be overcome if the water content of one compartment or the averaged water content of the cells can be measured independently, for instance by interference microscopy in the case of isolated cells or suspensions (von Zglinicki et al. 1987) or by bulk specimen X-ray microanalysis (Zs-Nagy et al. 1982, von Zglinicki and Lustyik 1986). In the latter case also the knowledge of volume fractions of the cellular compartments in question is necessary (von Zglinicki and Bimmler 1987). However, by using this technique constant section thickness is required only over the short distance between adjacent compartments. This constancy was experimentally demonstrated (see above).

Conclusions

The requirements to the preparation of biological samples which ensure most probably unbiased X-ray microanalytical results at subcellular resolution can be fulfilled today. Among these requirements, the following appear to be critical:

The time elapsed after stopping the blood flow should be not more than about 10 sec, at least in oxygen-sensitive organs; trauma during sampling should be avoided or exerted for not more than a few milliseconds; and freezing should be fast enough to prevent osmotic water fluxes between compartments. Transfer of sections should be done by using a cryotransfer device (hydrated sections) or, in the case of frozen dried sections, they should be carbon-coated and at or slightly above room temperature. Measurements should be done with a cold stage in use, otherwise the measured concentrations of sulphur cannot be taken as unbiased.

Not only local elemental concentrations but also native intracellular water distributions can be estimated. With it, a wide range of applications of biological X-ray microanalysis is now open to study basic regulatory events at the level of single organelles in the native state.

References

Barer R, Dick DAT (1957)Interferometry and refractometry of cells in tissue culture. Exp Cell Res suppl 4:103-147

Cameron IL, Fullerton GD, Smith NKR (1988)Influence of cytomatrix proteins on water and on ions in cells. Scann Microsc 2:275-288

Cantino ME, Wilkinson LE, Goddard MK, Johnson ME (1986)Beam induced mass loss in hig resolution biological microanalysis. J Microsc 144:317-327

Chang SH, Mergner WJ, Pendergrass RE, Bulger RE, Berezesky IK, Trump BF (1980)A rapid method of cryofixation of tissues in situ for ultracryomicrotomy. J Histochem Cytochem 28:47-51

David E (1969)Die Nervenzelle unter differenten Sauerstoffmangelbedingungen. Doctor's thesis, Humboldt University, Berlin

Gupta BL, Hall TA (1981)The X-ray microanalysis of frozen-hydrated sections in scanning electron microscopy: an evaluation. Tissue Cell 13:623-643

Gupta BL, Hall TA (1982)Electron probe X-ray microanalysis. Techn Cell Physiol P128:1-52

Hagler HK, Lopez LE, Flores JS, Lundswick RJ, Buja LM (1983)Standards for quantitative energy dispersive X-ray microanalysis of biological cryosections: validation and application to studies of the myocardium. J Microsc 131:221-234

Hagler HK, Buja LM (1986)Effect of specimen preparation and section transfer techniques on the preservation of ultrastructure, lipids and elements in cryosections. J Microsc 141:311-317

Hall TA, Gupta BL (1982)Quantification for the X-ray microanalysis of cryosections. J Microsc 126:333-345

Linders PWJ, Stols ALH, Stadhouders AM (1981)Quantitative electron probe X-ray microanalysis of thin objects using an independent mass determination method.Micron 12:1-4

Rick R, Dorge A, Bauer R, Gehring K, Thurau K (1979)Quantification of electrolytes in freeze-dried cryosections by electron microprobe analysis.Scann Electr Microsc 2:619-625

Rick R, Dorge A, Thurau K (1982)Quantitative analysis of electrolytes in frozen dried sections. J Microsc 125:239-247

Robards AW, Sleytr UB (1985)Low temperature methods in biological electron microscopy, Elsevier, Amsterdam New York Oxford, p89

Saubermann AJ, Dobyan DC, Scheid VL, Bulger RE (1986)Rat renal papilla: comparison of two techniques for X-ray analysis. Kidney Int 29:675-681

Saubermann AJ, Heymann RV (1987) Quantitative digital X-ray imaging using frozen hydrated and frozen dried tissue sections. J Microsc 146:169-182

Somlyo AP, Somlyo AV, Shuman H (1979) Electron probe analysis of vascular smooth muscle. J Cell Biol 81:316-335

Zeitler E (1971) Trockengewichtsbestimmung von Teilchen mit dem Elektronenmikroskop. In: Schimmel G, Vogell W (eds) Methodensammlung der Elektronenmikroskopie. Wissenschaftliche Verlagsgesellschaft, Stuttgart

Zglinicki T von (1987) A mitochondrial membrane hypothesis of aging. J Theor Biol 127:127-132

Zglinicki T von (1988a) Intracellular water distribution and aging as examined by X-ray microanalysis. Scann Microsc 2:1791-1804

Zglinicki T von (1988b) Monovalent ions are spatially bound within the sarcomere. Gen Physiol Biophys 7:in press

Zglinicki T von, Bimmler M (1987) The intracellular distribution of ions and water in rat liver and heart muscle. J Microsc 146:77-85

Zglinicki T von, Bimmler M, Krause W (1987) Estimation of organelle water fractions from frozen-dried cryosections. J Microsc 146:67-75

Zglinicki T von, Bimmler M, Purz HJ (1986) Fast cryofixation technique for X-ray microanalysis. J Microsc 141:79-90

Zglinicki T von, Lustyik G (1986) Loss of water from heart muscle cells during aging of rats as measured by X-ray microanalysis. Arch Gerontol Geriatr 5:283-289

Zglinicki T von, Uhrik B (1988) X-ray microanalysis with continuous specimen cooling: is it necessary? J Microsc 150:in press

Zglinicki T von, Uhrik B, Zierold K (1988) X-ray microanalysis of frozen-dried cryosections: influence of cryotransfer and specimen cooling. In: Dickinson HG, Goodhew PJ (eds) EUREM 88, vol III. Institute of Physics, Bristol Philadelphia, p555

Zierold K (1986) The determination of wet weight concentrations of elements in freeze-dried cryosections from biological cells. Scann Electr Microsc 2:713-724

Zierold K (1988) X-ray microanalysis of freeze-dried and frozen-hydrated cryosections. J Electr Microsc Techn 9:65-82

Zs-Nagy I, Lustyik G, Bertoni-Freddari C (1982) Intracellular water and dry mass content as measured in bulk specimens by energy-dispersive X-ray microanalysis. Tissue Cell 14:47-60

Conference Discussion

B.L. Gupta: (1) Mitochondria exist in several states of energy metabolism determined by the substrate levels (ADP/ATP ratios etc.) in the cytoplasm. Did you have these metabolites in your albumine model for isolated mitochondria? (2) Drying agents such as silica gel and molecular sieves are a very potent source of contamination for frozen-dried samples when stored before microanalysis. Did you use them?
Th. von Zglinicki: (1) Mitochondria were always in an energy-depleted state. No metabolites were added. (2) In preliminary experiments we obtained the same results. Therefore, we never stored sections over drying agents, but used them immediately after overnight freeze-drying.

R. Rick: I noted that your K concentrations per liter cell water are relatively low - at least when compared to other mammalian tissues. I wonder whether those low values may be due to an underestimation of the instrumental contribution to the white radiation. I think that in many cases the film measurement does not provide an accurate estimate of that component.
Th. von Zglinicki: The concentrations we measured in heart muscle are in fact at the lower border of the range considered as physiological. On the other hand, mean K concentrations in rat liver, for instance, were measured at the upper border of physiological values by using the same system. There seems to be no significant bias in our quantitation, therefore. Otherwise, I agree with your statement concerning the problems of estimating the extraneous background contribution. We are using the local film count and a scaled in-hole count for our correction. At present, we are trying to minimize the extraneous radiation by using a beam-limiting aperture and slot grids.

J. Shelburne: Have you used your cryo-biopsy-needle for human tissues? What types of biopsies are especially well suited for this device?
Th. von Zglinicki: It is impossible to use that technique for studies in humans, because there is severe freezing injury to

the tissue surrounding the needle tip, too. It has only been used in animals, therefore.

The Subcellular Accumulation of Toxic Heavy Metals: Qualitative and Quantitative X-Ray Microanalysis

A.J. Morgan[1], N. Roos[2], J.E. Morgan[1], C. Winters [1],
[1]School of Pure and Applied Biology,
University of Wales College of Cardiff
P.O. Box 915,
Cardiff CF1 3TL, U.K.

INTRODUCTION

Several qualitative microprobe studies have been focused on the distribution of essential and non-essential metals between the various types of metal-rich invertebrate spherites (Brown, 1982; Morgan, 1984; Morgan and Winters, 1987). The conclusion drawn from these observations is that, in general and despite the competitive tendancy extant in multi-metal and multi-ligand situations *in vivo*, the metals are preferentially bound to cellular ligands according to their equilibrium constants for metal-ligand complexes (Nieboer and Richardson, 1980): Class A ('hard acid') metals such as Ca, Mg, Mn, Sr and Pb show a preference for oxygen donating ligands according to the sequence O>N>S; Class B ('soft acid') metals such as Cd, Cu, Hg conform to the ligand sequence S>N>O (Simkiss, 1981; Simkiss and Taylor, 1981; Simkiss et al., 1982; Morgan and Winters, 1987).

It is only relatively recently that fully quantitative microprobe techniques have been employed: (a) to assess the utility of analysing metal sequestering spherites *in situ* for pollution biomonitoring purposes (Morgan and Morgan, 1988a); and (b) to examine in detail the binding of extraneous Zn (Morgan and Morgan, 1988b) and Pb (Morgan and Morgan, 1988c) within the matrix of certain earthworm spherites. The biological outcome of these studies are not germane in the present context, but three technical considerations may be of some general microprobe moment. First, to achieve quantitative analyses of heavy metals good reliable

[2]E.M.L.B. University of Oslo, P.O. Box 1062, Blindern, Oslo, Norway

standards are required; this paper describes our provisional, and very promising, attempts to combine the standard approach of analysing thin sectioned standards with a ratio technique for the rapid acquisition of the necessary standard conversion factors (Morgan and Winters, 1988a). Second, energy spectrometry provides multi-element information, and yet the data is seldom (Jones et al, 1986, 1988) analysed by multivariate statistical procedures. We present examples of the use of partial correlation to determine the relationship between two element-variables within spherite populations and, most pertinently, to extract the influences of co-variables on the relationship.

Observations

1. Standards for the quantitation of heavier elements in thin specimens by the Hall Continuum Normalisation Method.

Standardless quantitative X-ray microanalysis of biological specimens is impossible (Roomans, 1988). Methods for the transformation of observed X-ray intensities in thin films to actual element concentrations have been available for some time (Morgan, 1985). Similarly, several different types of thin standards have been formulated from which the conversion factors necessary to facilitate the transformations can be derived (Roomans, 1979). The fact that new standards are still being introduced and assessed. (Roos and Barnard, 1984; Roos and Morgan, 1985; Lupton and Saubermann, 1986) suggests that the ideal standard has yet to be formulated. The reasons for this are probably varied, but probably include the desire for a simple preparative protocol and good long-term storage properties, the necessity for the matrix to permit the homogeneous distribution of element(s) and to be stable under realistic electron irradiation conditions.

In our laboratory we have opted for sectioned aminoplastic standards, produced in most essential respects (Morgan and Winters, 1988b) according to the original methods of Roos and Barnard (1984). These aqueous standards possess many attractive properties (Roos and Morgan, 1985). Stimulated initially by the need for a wide range of standard factors, for light through to very heavy elements, we sought a means of avoiding the sectioning of a standard (in practise 4 or 5 standards of different concentration) for each element of interest. The approach adopted, which will be described in full elsewhere, is firstly to determine a reliable conversion factor for a given element dispersed in aminoplastic. The element chosen was calcium, but the choice is not critical providing there is a salt available which possesses good aqueous solubility properties and does not precipitate out of solution during the spontaneous polymerization of the glutarldehyde urea (aminoplastic) mixture. Having determined a 'reference standard factor' (i.e. for Ca), the standard factor for any other analysable element can be calculated from its previously established overall sensitivity relative to the reference element. The technique seems to be equally applicable to K- and L- line series, and in principle is not novel. Hall and Gupta (1983), for example, indicated that providing the concentration of one element in a sample is already known, the concentrations of all others can be determined by the 'ratio method' from relative sensitivity data. The outline theory is as follows:

The Hall Continuum Method (Roomans, 1988) indicates that for fixed analytical conditions the concentration (C_i) of element i in a thin specimen can be derived from the following equation; providing an absolute thin standard containing element i is available:

$$C_{i_{spec}} = C_{i_{std}} \cdot \frac{(P_i/W)_{spec}}{(P_i/W)_{std}} \cdot \frac{G_{spec}}{G_{std}} \qquad (1)$$

(where: P = observed peak intensity; W = intensity of the continuum radiation; G = $\overline{Z^2/A}$; spec = specimen; std = standard).

But a standard factor (conveniently disignated the FST_i factor) can be determined from the absolute standard, a sectioned aminoplastic say (Roos and Barnard, 1984; Roos and Morgan, 1985), by the application of equation 2:

$$FSTi = \frac{(P_i/W)_{std}}{C_{i\ std}} \cdot G_{std} \qquad (2)$$

Clearly, by the substitution of equation 2, equation 1 can now be simplified to:

$$C_i = \frac{(P_i/W)_{spec} \cdot G_{spec}}{FST_{i\ std}} \tag{3}$$

Before this equation can be applied for the determination of the concentrations (expressed usually as mM Kg^{-1} dry or wet weight, depending on the state of the specimen at the time of analysis) of each detectable element in the sample a complete suite of FST values must be known.

In a hypothetical thin film standard containing equimolar concentrations of two elements, x and y, the ratio of their respective FST's is represented as:

$$\frac{FST_x}{FST_y} = \left[\frac{\left(\dfrac{(P_x/W) \cdot G}{C_x}\right)}{\left(\dfrac{(P_y/W) \cdot G}{C_y}\right)}\right] \tag{4}$$

Equation 4 can be much simplified because in this example $C_x = C_y$. Thus:

$$FST_x = \left(\frac{P_x}{P_y}\right) \cdot FST_y \tag{5}$$

Since the ratio of the peak intensities (P_x/P_y) of two equimolar elements is an expression of their overall relative sensitivities under given analytical conditions, equation 5 becomes:

$$FST_x = FST_y \cdot \text{relative sensitivities (xy)} \tag{6}$$

In fact, Roos and Bernard (1984) implicity support this relationship because they demonstrate that FST values for given elements are a

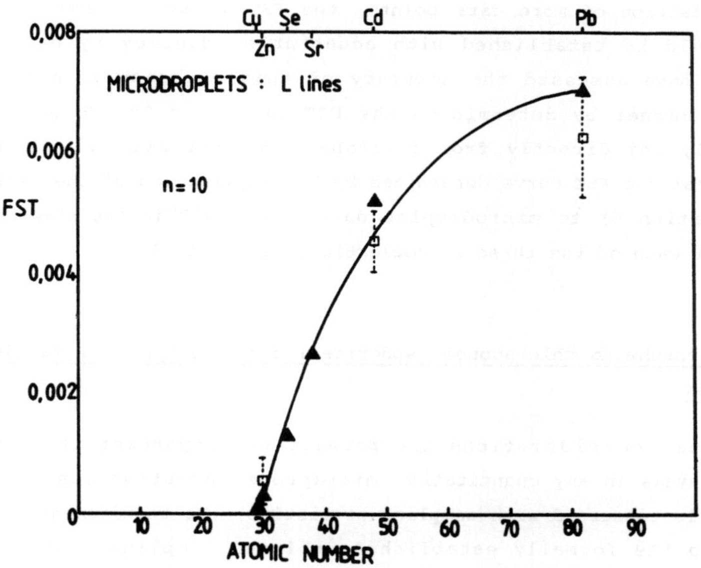

Fig.1. FST values (L-lines) for the elements Cu, Zn, Se, Sr, Cd and Pb determined by the 'ratio method' from the 'reference FST' of Ca, and plotted against atomic number (▲——▲). The FST values (mean ± 1 S.D) for Zn, Cd and Pb (L-lines) determined directly from sectioned aminoplastic standards (n=4) are included for comparison. Note that the hand drawn FST 'ratio curve' falls within the standard deviations of the aminoplastic values.

measure of the efficiency of a detector system for those elements.

To date we have used equation 6 to determine the FST values for the L-lines of Cu, Zn, Se, Sr, Cd and Pb (entered as FST_x) from the reference FST of Ca (entered as FST_y) already determined from sectioned aminoplastics. The sensitivities of each element relative to Ca were determined from the analysis of sprayed microdroplets of equimolar solutions (Morgan, Davies and Erasmus, 1975). Two practical points are worth underlining. First, the relative sensitivities could equally well be determined from the anlaysis of microcrystals of well characterised minerals. Second, it is important that the aminoplastic standard and the binary standards are analysed under identical conditions.

The results of this exercise are encouraging (Fig.1). The FST values of the L-lines of elements Cu (Z=29) to Pb(Z=82) form a smooth curve. The implications of this is that when the curve becomes more reliable with the aquisition of more data points, the FST values of unmeasured elements could be established with acceptable accuracy by direct reading. We have assessed the accuracy of this approach in a very provisional manner by determining the FST values of Zn, Cd and Pb independently and directly from sectioned aminoplastics. It is noteworthy that the FST curve determined by the application of the ratio method (equation 6) to microdroplet data falls within the standard deviations of each of the three aminoplastic values (Fig.1).

2. Metals in earthworm chloragosome spherites; data analysis by partial correlation

Statistical considerations are potentially important at three procedural levels in any quantitative microprobe investigation. The first level is concerned with sampling statistics; the requirements are analogous to the formally established unbiased sampling protocols prevalent in morphometry and stereology. The second level concerns the need to collect raw X-ray energy spectra with sufficiently good count statistics to permit the accurate deconvolution of overlapping peaks, and the measurement of peak areas. The third level concerns the statistical treatment of the multi-element quantitative data obtained from one or more cellular domains. Certain aspects of this third and neglected consideration will be dwelt on briefly here.

The earthworm possesses an unusual tissue, the chloragogenous tissue, located on the coelomic side of the alimentary canal. Qualitative and semi-quantitative microprobe analyses of air-dried smears of the chloragogenous tissue of earthworms inhabiting Pb, Zn, and Cd - polluted soils indicate that the chlorogocytes contain at least two distinct metal-squestering vesicles or spherites; an ovoid, very numerous, non-inducible Ca, Zn, Pb and P accumulating spherite called the chloragosome; a less numerous, less electron-opaque, inducible Cd and S, accumulating spherite, called the cadmosome (Morgan and Morris, 1982; Morgan, 1984). Circumstantial evidence, based on our recent biochemical observations (Morgan et al., 1988), suggests that the cadmosome contains a complex of Cd and a low molecular weight, cysteine-rich, metallothionein - like protein.

Fig.2. Thin cryosection of the chloragogenous tissue in the posterior alimentary region of the earthworm, Lumbricus rubellus, collected from highly contaminated (Pb, Zn and Cd) soil. The fresh tissue was frozen with the 'Hagler Cold Pliers' method (Hagler and Buja, 1984); mounted in a vice-type chuck and sectioned at a specimen temperature of about 130K; hydrated sections transferred to coated 75 mesh Cu grids; grids transferred to the microscope on a Gatan cold-stage, and the sections freeze-dried at 193K under the microscope vacuum. Analysis (see Table 1) was performed at 80 kV in the STEM mode in a JEOL 100 CX equipped with a LINK 860 Series II ED system and QUANTEM/FLS software. Note the dense granules in the chloragogenous tissue (ch) and the gut epithelial layer (e).

Earthworms tend to accumulate heavy metals from contaminated soils (Morgan and Morgan, 1988a), a large proportion of which are sequestered within the chloragogenous region (Morgan and Morgan, 1988d). Examination of thin cryosections reveals the presence of numerous dense inclusions within the chloragog of worms sampled from polluted soils (Fig.2), although it is difficult to distinguish different granule types

Table 1: Composition (mM kg^{-1} dry weight) of the spherites in cryosectioned chloragogenous tissue of the earthworm, *Lumbricus rubellus* inhabiting polluted soil.

	Type I (chloragosome)	Type II (chloragosome)	Type III (residual bodies)	Type IV (Cadmosome)
Mg	41 ± 37	O	48 ± 46	O
P	988 ± 130	1336 ± 155	141 ± 64	150 ± 63
S	613 ± 264	714 ± 197	1969 ± 280	1419 ± 347
K	49 ± 24	24 ± 11	57 ± 40	107 ± 25
Ca	619 ± 81	715 ± 94	201 ± 67	110 ± 88
Cd	15 ± 18	8 ± 6	834 ± 204	418 ± 100
Pb	47 ± 46	301 ± 76	314 ± 71	72 ± 0.7
(Zn)	(Yes)	(Yes)	(-)	(-)

on the sole basis of morphological criteria. However, the unbiased quantitative analysis of most of the dense structures within a field such as that depicted in Fig.2 yields information that may allow the definition of up to four spherite types (Table 1). This subdivision, albeit subjective and non-statistical, is based on composition information only. Moreover, it is based on a consideration of the entire spectrum of elements present in each analysed compartment. Obviously, such a 'fingerprinting' approach would be much more powerful if performed by a multivariate statistical procedure, such as the one used by Jones et. al. (1986, 1988) to discriminate between chemically distinct cohorts of erythrocytes during storage. The general point to be made is that microprobe data are usually presented and interpreted as mean values for each individual element in turn. This approach ensures that variations, especially those expressed by a small cohort of functionally influential organelles or cells, are overlooked because they are subsumed by the deviation around the population mean.

TABLE 2:
Coefficients of linear and partial correlation of element relationships within individual chloragosomes* of L. rubellus from two metalliferous mine sites.

	LINEAR CORRELATION			PARTIAL CORRELATION			
	P: Zn	Ca: Zn	S: Zn	Zn: P (-Ca)	Zn: P (-S)	Zn: Ca (-P)	Zn: Ca (-S)
Llantrisant (n=60)	0.955 (P<0.001)	0.747 (P<0.001)	0.935 (P<0.001)	0.932 (P<0.001)	0.662 (P<0.001)	-0.578 (P<0.001)	-0.109 (N.S)
Wemyss (n=33)	P: Ca	P: Pb	Ca: Pb	P: Pb (-Ca)	Ca: Pb (-P)		
	0.835 (P<0.001)	0.897 (P<0.001)	0.635 (P<0.001)	0.872 (P<0.001)	-0.506 (P<0.01)		

The chloragosomes were prepared for analysis by smearing and air-drying the chloragogenous tissue on e.m. grids (Winters & Morgan, 1988). Element concentrations, expressed as mM kg^{-1} dry weight, were determined by LINK QUANTEM FLS software.

Multivariate statistical procedures can not only discriminate between compartments, but they can facilitate the examination of the actual relationship between any two variables within a compartment by extracting the influence of other variables on the relationship. We have, for example, been able to enhance our understanding of the relationship between the major element constituents of earthworm chloragosomes by analysing the statistical data by partial correlation (Bailey, 1981). The benefits of this approach are summarised in Table 2. In the Llantrisant worms, linear correlation analysis suggests that there is a significant positive correlation between Ca and Zn in the chloragosome matrix. However, the subtraction of the influences of both

P and S (both variables being very strongly correlated with Zn) by partial correlation suggests that the correlation between Ca and Zn is negative. Similarly, in the Wemyss chloragosomes the apparent positive (linear) correlation betwen Ca and Pb is reversed to a significant negative Ca: Pb relationship when the influence of the P co-variable is extracted (Table 2). This implies that Pb is sequestered by exchange for Ca within the chloragosome matrix, and is probably bound to a P-rich ligand. An analogous Ca-Pb exchange occurs in bone mineral.

Acknowledgements

We thank Ulla Kinde for assistance with the cryomicrotomy, Mr. V.T. Williams for photographic assistance, Mrs Annette Robson for typing the manuscript.

References

Bailey NTJ (1981) Statistical methods in biology, 2nd edn. Hodder and Stoughton, London.

Brown BE (1982). The form and function of metal-containing 'granules' in invertebrate tissues. Biol Rev 57: 621-667.

Hagler HK Buja LM (1984) New techniques for the preparation of thin freeze dried cryosections for X-ray microanalysis. In: Revel JP Barnard T Haggis GH (eds) Science of biological specimen preparation Scanning Electron Microscopy Inc., AMF O'Hare, Chicago, p 161

Jones HB Morgan AJ Lovell DP Bellamy D. (1986) Statistical analyses of multielement electronprobe data correlated with morphological changes in stored human erythrocytes. Micron Microscopica Acta 17: 307-317.

Jones HB Morgan AJ Lovell DP Bellamy D (1988). Identification of chemically-unique individual cells of an erythrocyte population during storage in Acid Citrate Dextrose: analysis of electronprobe data by analysis of variance. Proc Roy Microsc Soc 23/4: 212

Lupton JB Saubermann AJ (1986) A new and simple method for preparation of thin aminoplastic standards for X-ray microanalysis. J. Microsc., 144: RP3-4.

Morgan AJ (1984) The localization of heavy metals in the tissues of terrestrial invertebrates by electron microprobe X-ray analysis. Scanning Electron Microsc 1984/IV: 1847-1865.

Morgan AJ (1985) X-ray microanalysis in electron microscopy for biologists. Oxford University Press/Royal Microscopical Society, Oxford.

Morgan AJ Davies T Erasmus DA (1975) Analysis of droplets from iso-atomic solutions as a means of calibrating a transmission electron analytical microscope (TEAM). J. Microsc 104: 271-280.

Morgan AJ Morris B (1982) The accumulation and intracellular compartmentation of cadmium, lead, zinc and calcium in two earthworms species (Dendrobaena rubida and Lumbricus rubellus) living in highly contaminated soil. Histochemistry 75: 269-285.

Morgan AJ Winters C (1987). The contribution of electron probe X-ray microanalysis (EPXMA) to pollution studies. Scanning Microscopy 1: 133-157.

Morgan AJ Winters C (1988a) Standards for the quantification of light and heavy elements by the continuum normalisation method: thin aminoplastic sections and sprayed microdroplets. Proc Roy Microsc Soc 23/4: 207-208

Morgan AJ Winters C (1988b). Practical notes on the production of thin aminoplastic standards for quantitative X-ray microanalysis.

Morgan JE Morgan AJ (1988a) Earthworms as biological monitors of cadmium, copper, lead and zinc in metalliferous soils. Environ Pollut 54: 123-138.

Morgan JE Morgan AJ (1988b) Zinc sequestration by earthworms (Annelida: Oligochaeta) chloragocytes. An in vivo investigation using fully quantitative electron probe X-ray micro-analysis. Histochemistry (in press).

Morgan JE Morgan AJ (1988c submitted). The effect of lead incorporation on the elemental composition of earthworm (Annelida: Oligochaeta) chloragosome granules.

Morgan JE Morgan AJ (1988d) Calcium-lead interactions involving earthworms. Part 2: The effect of accumulated lead on endogenous calcium in Lumbricus rubellus. Environ Pollut 55: (in press).

Morgan JE Norey CG Morgan AJ and Kay J (1988) A comparison of the cadmium-binding proteins isolated from the posterior alimentary canal of the earthworms <u>Dendrodrilus rubidus</u> and <u>Lumbricus rubellus</u>. Comp Biochem Physiol (in press).

Nieboer E Richardson DHS (1980). The replacement of the non-descript term 'heavy metal' by a biologically and chemically significant classification of metal ions. Environ Pollut (Ser B) 1: 3-26.

Roomans GM (1979) Standards for X-ray microanalysis of biological specimens. Scan Electron Microsc/1980/II: 309-320.

Roomans GM (1988) Quantitative X-ray microanalysis of biological specimens. J Electron Microsc Tech 9: 19-43.

Roos N Barnard T (1984) Aminoplastic standards of quantitative X-ray microanalysis of thin sections of plastic-embedded biological material. Ultramicroscopy 15: 277-286.

Roos N Morgan AJ (1985). Aminoplastic standard for electron probe X-ray microanalysis (EPXMA) of ultrathin frozen-dried cryosections. J Microsc 140: RP3-4.

Simkiss K (1976) Intracellular and extracellular routes in biomineralisation Symp Soc Exp Biol 30: 423-444.

Simkiss K (1981) Cellular discrimination processes in metal accumulating cells J exp Biol 94: 317-347.

Simkiss K Jenkins KGA McLellan J Wheeler E (1982) Methods of metal incorporation into intracellular granules. Experientia 38: 333-335.

Simkiss K Taylor M. (1981) Cellular mechanisms of metal ion detoxification and some new indices of pollution. Aquatic Toxicol 1: 279-290.

Somlyo AP Bond M Somlyo AV (1985) Calcium content of mitochondria and endoplasmic reticulum in liver frozen rapidly <u>in vivo</u>. Nature 314: 622-625.

Winters C Morgan AJ (1988) Quantitative electron probe X-ray microanalysis of lead-sequestering organelles in earthworms: technical appraisal of air-dried smears and freeze dried cryosections. Scanning Microscopy 2: 947-958.

Zierold K (1988) X-ray microanalysis of freeze-dried and frozen-hydrated cryosections. J. Electron Microsc. Tech 9: 65-82.

Conference Discussion

K. Zierold: I agree with your arguments to use multivariate statistics to evaluate X-ray microanalytical data. However, by this "finger printing" procedure you risk to find high correlation coefficients by preparation-induced artefacts, for example, in precipitates. Please comment!

J.A. Morgan: I presume that the thrust of your comment is that the co-precipitation of two or more elements, for example Ca and P, during specimen preparation will inevitably lead to a statistically strong, but conceivably entirely spurious, positive correlation between the variables in question. This is, of course, true but does presuppose a major translocation of elements between major sub-cellular compartments during specimen preparation. My comments about element-element correlations, and particularly my admonition concerning the indiscriminate use of linear correlation analysis, were meant to apply exclusively to preparations free of such serious artefacts - these are the only biologically interesting specimens, in any case. [Incidentally, it is quite feasible to 'derive' a spurious negative correlation between a very heavy element whose concentration is very variable, and a very light element whose concentration is fairly constant, due to uncorrected absorption of the weak X-rays from the light element]. However, your comment refers to my use of the term chemical 'finger printing'. I use this evocative term in a similar way to Appleton (1978; In: Electron probe microanalysis in biology, ed. D.A. Erasmus, pp. 148-182, Chapman & Hall, London), and it is meant to suggest that the entire quantitatively measured elemental composition data are used (a) to determine the heterogeneity within a population of similar structural compartments, and (b) to compare different populations of similar compartments. It is very significant that Hagler et al. (this volume) have identified a pathophysiologically important heterogeneity in a cultured population of rat myocytes on the basis of analytical data for one element (- calcium -) only. Factorial and classification statistical methods are available for handling multivariate data; we referred to the papers by

Jones et al. (1986, 1988), but others are available (e.g. Ballan-Dufrancais et al., 1985 - Biol. Cell. 53:283-292). The multi-spectral imaging and scatter diagram construction procedures, recently described by El Gamati et al. (J. Microsc. 147, 149-158, 1987) in the context of scanning Auger microscopy, are also of interest. In concluding, I must return to a point implicit in your comment which emphasises that the product of any statistical analysis will only be amenable to valid biological interpretation if the original raw data were free from major inaccuracies and inconsistencies arising from preparative and analytical sources.

X-Ray Microanalysis of Cryosections Using Image Analysis

Albert J. Saubermann, M.D.
Dept. of Anesthesiology
Health Sciences Center
State Univ. of New York at Stony Brook
Stony Brook, New York 11794-8480, USA

It is possible for electron probe x-ray microanalysis to be used to generate quantitative digital images of elemental distribution in cryosections of biological specimens (Gorlen et al. 1984, Ingram et al. 1987, Saubermann and Heyman 1987, Fiori et al. 1988, Johnson et al. 1988). By combining low temperature specimen preparation methods with computerized electron beam placement, an elemental map can be produced which gives information in a readily interpretable and understandable manner. By using well established electron probe x-ray analytical algorithms for quantitation of x-ray signals obtained through energy dispersive systems, each pixel can yield valuable quantitative information. When many pixels are combined, the statistical power of this form of analysis can be appreciated. However, and perhaps even more importantly, when grey level values are assigned to a range of pixel values in forming a map, it is possible for the eye and brain to do the statistical processing in an intuitive fashion. Since information on several elements can be acquired simultaneously, it is also possible to superimpose, and, thereby, relate elemental concentrations to each other as well as to anatomic structure. When quantitated, elemental mass is related to wet weight mass, as well as to dry weight mass. By using frozen hydrated cryosectioning techniques and analysis methods, special elemental relationships often become apparent (Saubermann and Stockton 1988, Saubermann and Heyman 1987). However, direct x-ray imaging of frozen hydrated tissue sections presents a formidable list of problems. Among these problems are 1) low peak to background ratios of biological elements normally found in small wet weight concentrations; 2) indistinguishable morphology due to the lack of contrast making it very difficult, if not impossible, to identify an area to be mapped; and 3) a specimen which is highly radiation sensitive such that water loss occurs at probe currents necessary to permit statistically acceptable x-ray generation rates. The methods which we have developed, use an algorithm which basically circumvents these major problems. Because changes in dry weight mass fraction of elements may not reflect changes in wet weight mas fractions (since cell water content can change for various physiological and pathological reasons), it may be important to determine and appreciate both wet weight and dry weight mass fractions in a biological specimen (Saubermann and Stockton, 1988).

METHODS

Specimen preparation: Rapidly frozen tissue samples can be obtained by a variety of methods. Samples are placed on wooden pegs and plunged into melting Freon (Saubermann and Scheid, 1985) or *in situ* freezing using a pair of liquid nitrogen cooled, copper-jawed pliers is applied (Ingram and Ingram 1984, LoPachin et al. 1988). After freezing, the frozen specimen is mounted in a precooled vise-type three-point or two-point holder for cryosectioning. Cryosections (nominal thickness 0.5 μm) are then cut at -55°C, which minimizes cutting work and maximizes chip (section) smoothness (Saubermann et al. 1981a,b; Saubermann and Scheid 1985, LoPachin et al. 1988). A specially designed cryochamber (Burlington Scientific) which fits a Sorvall-MT2B microtome, is used for cryosectioning. Frozen hydrated sections are transferred to a special beryllium specimen holder consisting of a beryllium grid covered with a carbon coated nylon film on the hollow holder (Saubermann et al. 1981a, b). The specimen holder, mounted in a beryllium-grooved transporter (specimen holder and cold stage interface), is precooled within a special chamber in the microtome before a cryosection is transferred to the holder's surface. The specimen holder assembly is then transferred to the cold stage in the vacuum column of an AMRay 1400T scanning electron microscope using a specially designed transfer device (Saubermann et al. 1981a,b). This transfer device consists of a cooled copper heat sink in a vacuum-tight Delrin cylinder fitted with a vacuum valve allowing the cylinder to be evacuated once the specimen holder is withdrawn into the chamber prior to transfer to the scanning electron microscope stage (Saubermann et al. 1981a,b).

X-ray analysis: Spectra are collected using a Tracor Northern 2000 (Tracor Northern Inc.) energy-dispersive x-ray analysis system (EDS) with a 10 mm^2, 148 eV resolution Si(Li) detector. The detector's working distance is adjusted to less than 15 mm. A PDP 11/23 computer (Digital Corp.) with 128K memory and a 5.2 megabyte hard disk (DSD 880) is used for initial calculations and beam control. A Tracor Northern digital beam control system is interfaced with the scanning electron microscope to enable computer control over beam placement. A color monitor (Spectra Chrome) is used for initial tasks related to obtaining and processing elemental maps.

After the initial collection of data and preliminary processing, data files are transferred to an IBM AT/PC for image display and data retrieval. Thus the basic system is a conventional scanning electron microscope equipped with an EDS system using an integrated computer for beam control, initial spectra collection and processing (for a listing of the computer program used for data collection and initial analysis see Heyman and Saubermann, 1987). Imaging and data analysis is carried out on a conventional personal computer equipped with a professional graphics card and display system.

The basic analytical and display algorithm used is shown in Figs. 1a and b. Analyses are carried out in the hydrated state first to determine continuum generation rate for latter retrieval and calculation of hydrated mass fractions. After a suitable frozen hydrated section is identified and photographed, a measurement of hydrated wet weight mass is made from selected areas by determining the continuum generation rate between 4 and 6 keV. These data are obtained using 20 keV accelerating voltage and 0.4 nA probe current (measured in a Faraday cup on the cold stage) for 100 s live time with a rapid raster pattern so as to minimize dwell time at any single point to reduce radiation damage. The continuum generation rate is determined over an arbitrarily defined area by superimposing a grid pattern on the frozen hydrated section. Each area, depending upon magnification is usually less than 60 x 60 μm. Normally these area continuum count rates are obtained at a magnification of approximately 500x. The position of the area is marked on the photograph and the grid square's label is stored, along with the corrected continuum generation rate, in a computer file. This step is repeated until all suitable areas (defined as those areas where the cryosection is flat and not located over the Be grid bar) of the frozen hydrated section have been measured for continuum generation rate. Next, the electron beam is blanked off and the temperature of the cold stage is raised to permit the frozen hydrated section to freeze dry within the chamber vacuum at -60°C for 30 min. After drying is complete, the cold stage is again cooled and an area is chosen for imaging using morphological structures observed in the scanning transmission mode (STEM). An area for imaging is usually selected at magnifications of around 2000x so that the area generally fits completely within one of the "areas" previously used to obtain the continuum generation rate. Thus the particular area which is to be analyzed and imaged is matched to its corresponding hydrated area. An analysis matrix is then chosen (64 x 64, 128 x 128, or 256 x 256 pixels), and the beam positioned over the first pixel location. After first setting beam conditions (1.4 nA, 20 keV), spectra are then collected using a 4 second dwell time (live).

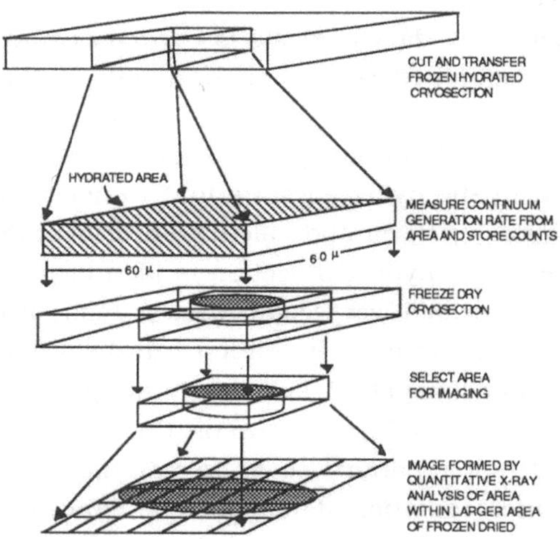

Fig. 1a

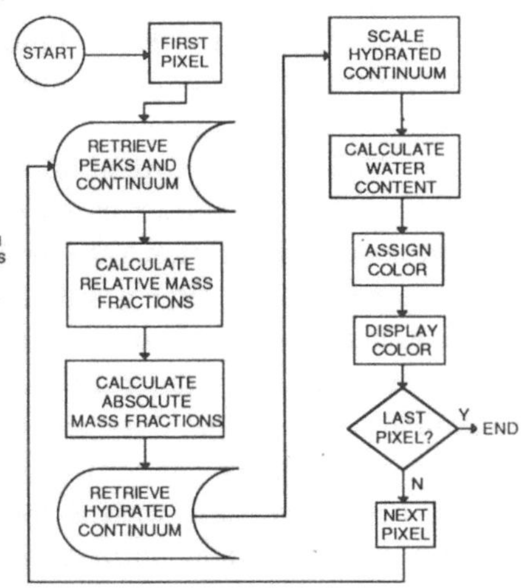

Fig. 1b

Fig. 1a) Proceedure used for data acquisition to produce quantitative x-ray maps. Fig. 1b) Flow diagram outling algorithm used for initial display of quantitative elemental maps.

During the 4 seconds, over which a spectrum is collected, the previous spectrum is corrected for extraneous background x-rays (generated mainly from the nylon support film) and a determination of the dry mass is made from continuum generation rate over the same energy band as in the hydrated state. The spectrum is then passed trough a digital band bass filter (top hat filter) and integrated peak counts are determined for characteristic x-rays from elements of interest. Seven standard elements are normally measured (Na, P, S, Cl, K, Ca, and Mg). These elemental peak count rates and continuum count rates are stored in a separate file along with the hydrated continuum generation rate for the larger area.

To display an elemental image or map, the x-ray generation rate from the hydrated section (obtained at 0.4 nA probe current for 100 seconds live time) is mathematically scaled to a generation rate at 1.4 nA for 4 seconds live time. This calculation is based upon the known linear relationship between continuum generation rate and probe current:

$$W_S = W_H \frac{T_A}{T_H} \frac{I_A}{I_H} \tag{1}$$

where W_S and W_H are continuum counts scaled and from the hydrated area respectively, T_A is acquire time (live) for pixel analysis, T_H is acquire time for hydrated areas, I_A is probe current used for pixel analysis, and I_H is probe current for hydrated area measured. The scaled continuum counts for the hydrated state are used to calculate wet weight concentrations and pixel water content. Wet weight concentration and water content (percent water) are calculated pixel by pixel using the following equations:

$$C_{H_2O} = \left[1 - \frac{W_D}{W_H}\right] * 100 \tag{2}$$

Absolute mass fractions:

$$C_x = R_{x\ \mathrm{Dried}}\ S_x \tag{3}$$

and

$$C'_x = R_{x\ \mathrm{Dried}} \left[\frac{100 - C_{H_2O}}{100} \right] \tag{4}$$

where W_D and W_H is the corrected continuum counts from the dried and hydrated section respectively, C and C' are absolute elemental concentrations expressed as mmol/kg dry or wet weight respectively, S is a calibration factor experimentally determined for element x (standardization) and R is the ratio of peak x-ray counts for element x (minus background) to net continuum counts (minus background).

Image display: Data files containing the pixel values are transferred to an IBM AT/PC configured with Professional Graphics Display (IBM) monitor and Professional Graphics Controller, a 40 megabyte hard disk and two 5 1/4" inch high density disk drives. Four megabytes of core memory are configured as 3 megabytes of virtual disk and 1 megabyte as core memory. Imaging programs are written in Fortran using Fortran callable graphic routines using the Graphic Kernal System (GKS).

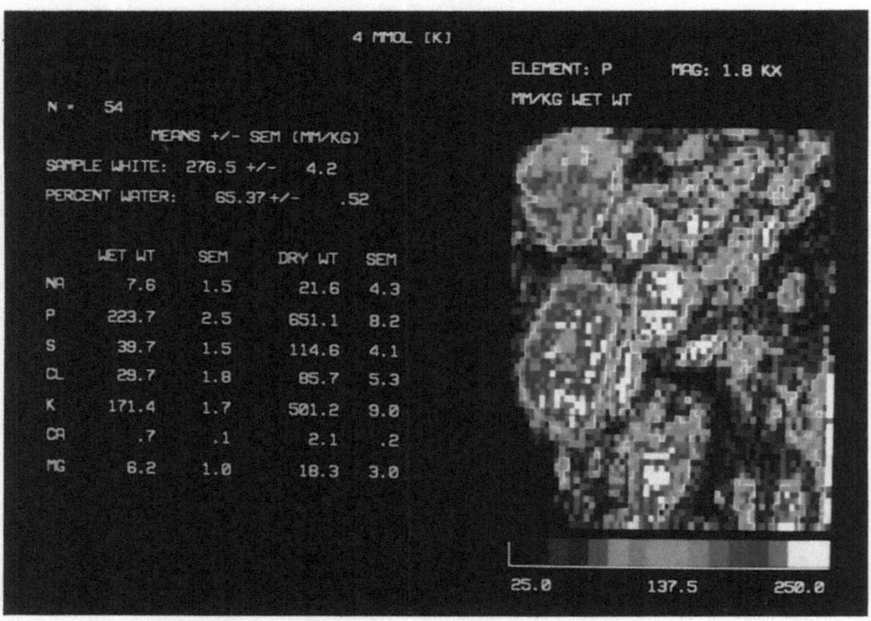

Fig.2 Quantitative digital x-ray image of leech ganglia shown as a pseudo-color map (16 intensity levels) of phosporous (mmol/Kg wet weight). Quantitative information, retrieved from pixels from the area outlined in black, is displayed as both wet and dry weight mass fractions. Image is 64 x 64 pixels.

RESULTS

Quantitative digital x-ray images using elemental mapping methods are shown in Fig. 2. Each pixel in these images represents a spatial resolution approximated by beam diameter (~120 nm). Maximum pixel density/analytical area is basically limited by beam diameter, by section thickness, and by electron scattering. These images show a good correlation between anatomic morphology and elemental location. Individual pixels have elemental quantities which are equivalent to values obtained by using conventional operator placement of beam position and single point analysis methods. Both mmol/kg wet weight and dried weight can be obtained with good precision and accuracy (Saubermann and Heyman 1987). Values of elemental concentration and water content can be retrieved from these images for data retrieval and quantitative or statistical comparisons. These images (64 x 64 pixels) each required about 5-6 hours to acquire (4 s dwell time plus dead time correction).

DISCUSSION

Electron probe x-ray microanalysis of cryosectioned biological specimens combines morphological information with a powerful analytical method. Digital computer control of beam positioning has made it possible to form elemental maps using characteristic x-ray counts obtained with EDS systems (Gorlen et al, 1984; Fiori et al., 1985; Ingram et al., 1987; Saubermann & Heyman, 1987). While many commercially available systems can produce qualitative or semi-quantitative images, to produce an elemental map where each pixel represents a measured and quantified elemental mass fraction requires application of a suitable algorithm. Such an algorithm must do several things including: 1) suppress background counts; 2) extract characteristic x-ray peak counts; 3) deconvolute peak overlaps (such as the classic $K_{K\beta}$ - $Ca_{K\alpha}$ overlap); and 4) standardize the x-ray peak counts. Fiori et al. (1985) and Gorlen et al. (1984), for example, have very elegantly demonstrated the feasibility of such an approach using a very creative and sophisticated algorithm to provide quantitative chemical images of cellular structures. The intuitively interpretable qualities, inherent in elemental maps, can be appreciated from their images.

More recently we have demonstrated a method for application of the general principles of quantitative elemental mapping to the problem of analysis of frozen hydrated and then frozen dried cryosections. To do this required adaptation of our analytical algorithm using the Hall method of continuum normalization (Hall et al. 1973).

Frozen hydrated cryosections present a number of unique problems for elemental imaging. These problems include a specimen which is difficult to prepare and handle, a specimen which is radiation sensitive, a specimen in which little morphological information on cellular structure can be decerned until the water is removed, and a specimen where elemental mass fractions may be very low because of the water mass (i.e. low peak to background ratios). To overcome these problems we adapted our algorithm for conventional operator placed beam positioning to an x-ray imaging algorithm (Saubermann and Heyman 1987). The frozen hydrated cryosection is used to determine the continuum generation rate for the hydrated state from an area to be mapped, later, after the section is dried. This initial step, of measuring the continuum generation rate, is carried out at low current densities and long count rates (0.4 nA, 20 keV, 100 s). Previous experience has shown that radiation induced mass loss is analytically insignificant under these conditions (Saubermann & Scheid, 1985). By arbitrarily dividing a frozen hydrated cryosection into areas, it is possible to obtain a measure of continuum generation rate representing that area. This step is carried out blindly in the sense that the operator does not know if there will be any structures of interest in that area until the section is subsequently freeze dried and the underlying morphology is apparent. The second step, after freeze drying, is to begin a pixel by pixel acquisition of spectra from an area of interest using short dwell times and high current densities (1.4 nA, 20 keV, 4 s). Continuum generation rate in the dried state is compared with that of the hydrated state to measure pixel water content. In order to do this, we make the assumption that hydrated continuum generation rate in that, now dried, pixel was equivalent to that of the larger area previously acquired in the hydrated state. This assumption can be experimentally demonstrated to be true providing the specimen has a relatively homogeneous average atomic number/mass thickness (a reasonably valid assumption - see Hall and Gupta, 1982) and has a relatively constant thickness throughout the area. This later assumption is generally true; when it is not true, it is readily apparent in the x-ray maps (Saubermann and Heyman, 1987).

By collecting data from samples in the dried state and then converting the resulting information into both wet and dry weight mass fractions, many of the inherent problems of direct analysis of frozen hydrated cryosection are adverted. For example, by using the dried state and analyzing at low temperature it is possible to use relatively high current densities without analytically significant radiation induced mass loss to 0.5 μm thick sections. The actual limitation to probe current is detector dead time. Since higher probe current results in increased dead time, dwell time increases without additional advantage.

One of the major limitations of our algorithm is the need for relatively long times to acquire an image. During this time instrumental drift can be a problem and require compensation (Statham 1987). Gorlen et al. 1984, use more points acquired over a shorter time period to obtain a statistical sample of an area. They also use mathematical scaling for dead time correction. This method has the advantage of providing an image in a relatively short period of time. Our algorithm, trades acquisition time for individual pixel precision giving each pixel a higher degree of statistical importance by paying the price of increased counting time (Saubermann and Heyman 1987).

It is clear that quantitative digital elemental images provide a unique perspective on elemental analysis. By combining morphological information and quantitative chemical information, one can intuitively understand relationships between elemental content and anatomic form. More importantly each elemental image is a quantitative chemical micrograph analogous to the carbon, nitrogen, oxygen, etc "images" produced with standard transmission or scanning transmission electron microscopy which we generally consider as representing cellular morphology or anatomy. Studies of such images, using an experimental approach, has produced new understandings of the relationship between form and function. It is possible to consider, say, a potassium image, in the same way and, thus, relate chemical form to function. By comparing hydrated and dried images (water contents), elemental relationships and changes may be observed which could be overlooked using dried images alone. Furthermore, because the images are fully quantitative, one has the opportunity to pool data from many pixels to extract real numbers defining elemental mass fractions for comparison with similar areas under different conditions. Finally, the use of maps reduces operator bias. Such bias could inadvertently result, for example, if an operator places the electron beam over an area where elemental distribution appears strange or different, interprets the data as being anomalous, and discards that analysis.

The production of x-ray elemental images of frozen hydrated cryosections is a powerful analytical method for measuring the amount and location of water and other elements in biological specimens. These images provide easily interpretable information on the relationship of elements to structure, the relationship of elements to elements, and the relationship of elements to wet or dry mass. In combination with other experimental methods, x-ray mapping provides important new information on spatial relationships.

REFERENCES

Fiori C, Leapman R, Swyt C, Andrews B (1988) Quantitative x-ray mapping of biological cryosections. *Ultramicroscopy* 24:237-250.

Gorlen K, Barden L, Del Priore J, Fiori C, Gibson C, Leapman R (1984) Computerized analytical electron microscope for elemental imaging. *Rev. Sci. Instrum.* 55:912-921 .

Hall T, Anderson H, Appleton T (1973) The use of thin specimens for x-ray microanalysis in biology. *J. Microscop.* 99:177-182.

Hall T, Gupta B (1982) Quantification for the x-ray microanalysis of cryosections. *J. Microscop.* 126:333-345.

Ingram F, Ingram M (1984) Influences of freeze-drying and plastic-embedding on electrolyte distributions. In: *The Science of Biological Specimen Preparation.* J. P. Revel, T. Barnard, & B. Haggis, eds. SEM Inc., AMF O'Hare, Il. pp 167-174.

Ingram P, LeFurgey A, Davilla S, Lamvik M, Kopf D, Mandel L, Lieberman M (1987) Real-time quantitative elemental analysis and imaging in cells. *Analytical Electron Microscopy - 1987.* D. Joy, ed. San Francisco Press. S.F. Calif. Proceedings MAS meeting. pp 179-183.

Johnson D, Izutsu K, Cantino M, Wong J (1988) High spatial resolution spectroscopy in the elemental microanalyis and imaging of biological systems. *Ultramicroscopy* 24:221-236.

LoPachin R, Lowery J, Eichberg J, Kirkpatrick J, Cartwright J, Saubermann A (1988) Distribution of elements in rat peripheral axons and nerve cell bodies determined by x-ray microprobe analysis. *J. Neurochem.* 51:764-775.

Saubermann A, Beeuwkes R, Peters P (1981) Application of scanning electtron microscopy to x-ray analysis of frozen-hydrated sections. II. Analysis of standard solutions and artificial electrolyte gradients. *J. Cell Biol.* 88:269-273.

Saubermann A, Echlin P, Peters P, Beeuwkes R (1981) Application of scanning electron microscopy to x-ray analysis of frozen-hydrated sections. I. Specimen handling techniques. *J. Cell Biol.* 88:257-267.

Saubermann A, Heyman R (1987) Quantitative digital x-ray imaging using frozen hydrated and frozen dried tissue sections. *J. Microscop.* 146: 162-182.

Saubermann A, Scheid V (1985) Elemental composition and water content of neuron and glial cells in the central nervous system of the North American medicinal leech (*Macrobdella decora*). *J. Neurochem.* 44:825-834.

Saubermann A, Stockton, J (1988) Effects of increased extracellular K ln the elemental composition and water content of neuron and glial cells in leech central nervous system. *J. Neurochem* 51:1797-1807.

Statham P (1987): Quantitative digital mapping with drift compensation. *Analytical Electron Microscopy - 1987*. D. Joy ed. San Francisco Press, S.F. Calif Proceedings MAS Meeting. pp 187-190.

Conference Discussion

Th. von Zglinicki: During freeze-drying the specimen normally undergoes a shrinkage of about 10 to 20 %, that means, the density per area of dry mass is about 1.5 times higher in frozen-dried sections. Because you are using only background from frozen-hydrated sections, this should seriously bias your water content measurements.
A. Saubermann: Our linear shrinkage averages about 6 % and the shrinkage factor appears to be analytically insignificant in terms of our measurement of water content. This last statement is based upon direct measurements using comparative specimens and standards. The reason for this apparent insignificance of shrinkage on water content measurement in our specimens is unclear, however, it may be due in part to our use of thick (0.5 μm) sections where the errors induced by shrinkage may be of less significance.

P. Ingram: Do you think it reasonable to adopt (a) logical color scale(s) for display of quantitative concentration maps and if so, on what should it be based?
A. Saubermann: Since the use of colors help convey information on magnitude (quantity) and help differentiate one pixel from another, I strongly agree that it would be reasonable and highly desirable to have a logical uniform color scale. Obviously, there are many problems associated with color scales which include subjective ideas about colors as well as problems regarding photographic reproduction and color discrimination. Our color scale was the result of a "vote" taken in our laboratory to pick the most effective color scale of several. The final result was that it still has problems. Thermal scales have been suggested as the most effective and perhaps these scales should be considered more carefully.

A. Warley: Using your image analysis system is it possible to map the errors on the quantitation and, thus, give some idea on the probability of detecting low amounts of elements?
A. Saubermann: We have not yet written the computer program

which would permit us to map the errors. Having seen the work which Peter Statham has presented using error mapping, it is clear that there is much information contained in such maps which can be of value. Our data, used in construction of our images, will ultimately allow its use in error mapping and we plan to write the programs to do so.

Electron Probe X-Ray Microanalysis in the Silkmoth Antenna - Problems with Quantification in Ultrathin Cryosections

R.A. Steinbrecht and K. Zierold[1]

Max-Planck-Institut für Verhaltensphysiologie

D-8130 Seewiesen

F.R.G.

Introduction

The antenna of the male silkmoth has been a model system for olfactory physiology since the time when its response to the female sex pheromone could be recorded electrophysiologically (Schneider, 1957, 1984). The response triggers reproductive behaviour and is mediated by ten thousands of identical miniature sensory organs, the sensilla trichodea, each consisting of a slender cuticular hair which contains the sensory processes of two receptor cells surrounded by sensillum lymph. Three auxiliary cells surround the sensory cell bodies at the base of each hair and are supposed to play a key role in regulating the ionic composition of the sensillum lymph. Nevertheless, the morphology of olfactory sensilla is highly complex and its interpretation in terms of the electrical circuitry is far from clear (for review see Steinbrecht, 1987, De Kramer and Hemberger, 1987).

The silkmoth antenna, in addition, became a model system to study cryofixation and its various applications in functional morphology (Steinbrecht, 1984). The sensory hairs (length 100 µm, diameter 2 µm) and also the antennal branches (length 500-1500 µm, diameter 50-80 µm) are tiny cylindrical structures with a high surface-to-volume ratio optimally suited for plunge freezing (Fig. 1). Moreover, the cuticle provides optimal protection against desiccation and other damage during handling prior to freezing. Isolated anten-

[1]Max-Planck-Institut für Systemphysiologie, D-4600 Dortmund, F.R.G.

nae function normally in the electrophysiological experiment up to an hour and more and can be adequately stimulated immediately before cryofixation. Thus, the preparatory conditions are favourite for employing electron probe X-ray microanalysis (EPXMA) to study the electrolyte composition in the various compartments of the sensillum, and to compare the situation before and after stimulation.

Cryoembedding, cryosectioning and microanalysis

In our opinion, cryosectioning is the best preparation technique for the localization of mobile ions in cellular structures by EPXMA (Zierold and Steinbrecht, 1987). Cryoembedding in heptane (mp = - 90°C) was developed to transform the branched antennae with its fine hairs after freezing into a block suitable for sectioning in the cryoultramicrotome (Steinbrecht and Zierold, 1984). Sections of 70-100 nm thickness were obtained at -100° - -120°C in a Reichert FC4 cryoultramicrotome and transfered with an eyelash probe on Cu or Ni grids covered with a carbon-coated formvar film. After cryotransfer the sections were freeze-dried in the airlock of the electron microscope (Siemens ST100F) by disconnecting the cooling at a vacuum of $<10^{-4}$ Torr and allowing the specimen to reach temperatures above -80°C within about 30-60 min. The vapour pressure of heptane at these temperatures is more than hundred times higher than that of water so that it evaporates first and cannot interfere with the freeze-dried cryosections.

<u>Fig. 1.</u> Detail of the antenna of the silkmoth, *Bombyx mori*, showing the high surface-to-volume ratio of antennal branches and sensory hairs.

<u>Fig. 2.</u> Freeze-dried ultrathin cryosections of antennal branch (A, B) and sensory hair (C). The boxed area in A is shown in B at higher magnification. The various compartments of receptor cells (R) and auxiliary cells (A) can be clearly distinguished, because cell membranes are well visible, except where sectioned obliquely. Note the varying mass thickness in the extracellular sensillum lymph due to uneven sectioning and/or freeze-drying. Asterisks mark larger sectioning and/or freeze-drying artefacts. A auxiliary cell; C cuticle; D sensory dendrites; ER endoplasmatic reticulum; R receptor cell; SL sensillum lymph; SLS sensillum lymph space below hair base surrounded by microvilli and microlamellae of the auxiliary cell.

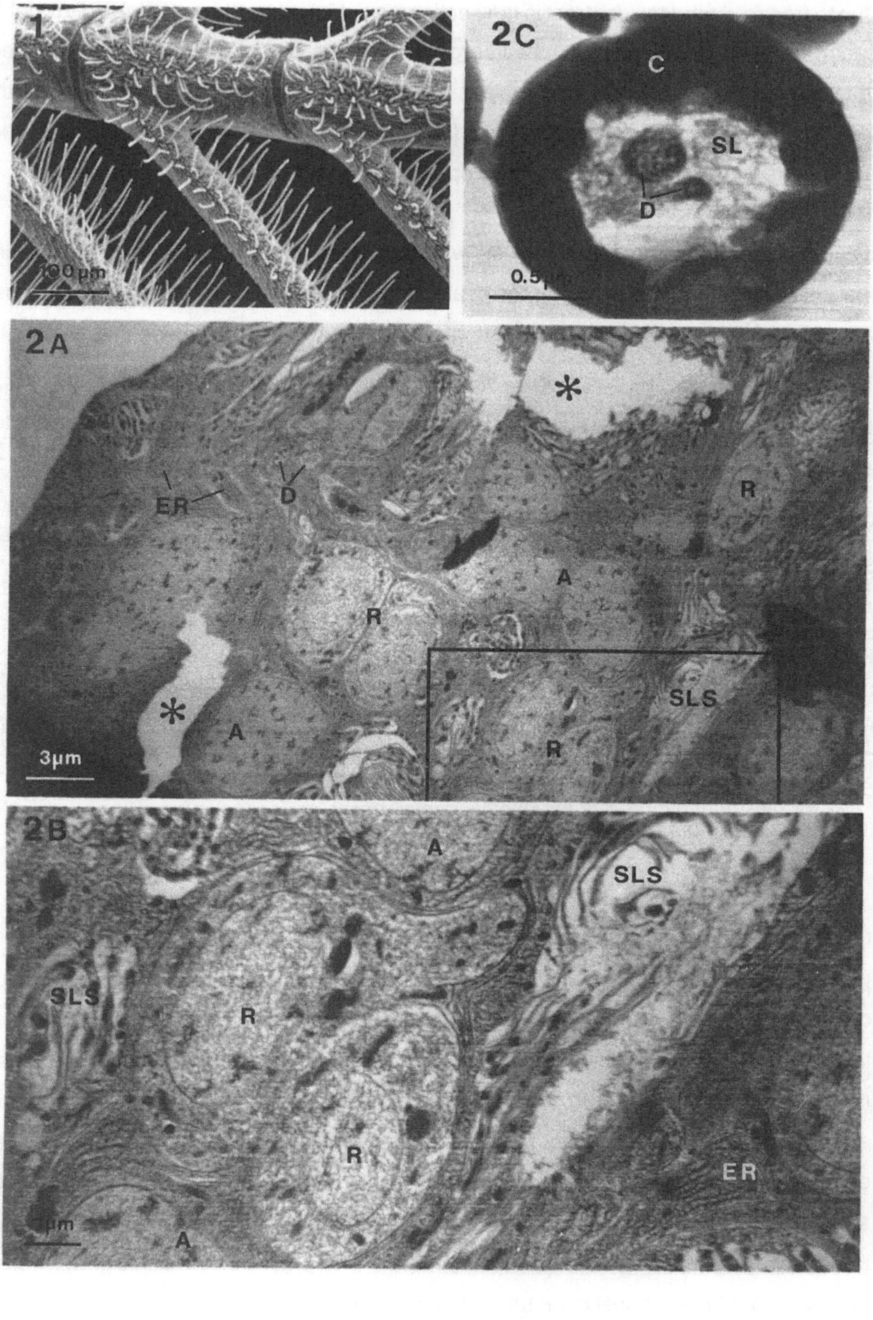
1
100 µm
2C
C
SL
D
0.5 µm
2A
*
ER
D
R
A
R
SLS
*
A
R
3µm
2B
A
SLS
SLS
R
R
ER
1µm
A

Due to freeze-drying the contrast in the sections is sufficient to allow unequivocal identification of all major compartments and organelles in the sensillum (Fig. 2A,B). Cryosections are routinely obtained from the tissue within the antennal branches, whereas the yield of sections of the fine sensory hairs (Fig. 2C) is much smaller. Probably, these separate more easily from the sectioned heptane and are then lost during transfer.

Energy dispersive X-ray microanalysis was performed using a SiLi-detector (Link) and the Link Quantem FLS program for evaluating the spectra (for analytical details see Zierold, 1988). A typical example of the results is given in Tab. 1. These results confirm

Table 1. Element distribution (expressed as mmoles/kg dry weight) in extra- and intracellular compartments of olfactory sensilla of *Bombyx mori* (data from Steinbrecht and Zierold, 1987, and unpubl.). Numbers in parentheses give standard deviation as % of the mean.

Compartment	n[1]	Na	Mg	P	S	Cl	K	Ca	ratio[2]
Sensillum lymph	10	n.s.	n.s.	n.s.	354 (24)	58 (22)	586 (34)	n.s.	1.62 (12)
Auxiliary cell cytoplasm	19	n.s.	66 (32)	239 (20)	198 (15)	22 (49)	337 (22)	n.s.	0.63 (11)
Receptor cell cytoplasm	26	n.s.	49 (41)	457 (20)	162 (16)	24 (23)	37 (20)	n.s.	0.86 (15)
Haemolymph	15	n.s.	137 (27)	454 (19)	157 (26)	97 (22)	96 (20)	36 (25)	0.21 (12)

[1] data from different sections (one animal)
[2] K/S for sensillum lymph, K/P for all other compartments
n.s. not significantly above detection limit (Na, Mg < 20, Ca < 10 mmoles/kg dry weight)

the high potassium content of sensillum lymph as has been measured in isolated droplets (Kaissling and Thorson, 1980). Moreover, preliminary data suggest that the potassium content of sensillum lymph is elevated after a strong stimulus, which would indicate an interaction between the sensory cells and the non-neural ion-transporting auxiliary cells (Steinbrecht and Zierold, 1987). The standard deviation of the concentration data, however, is very high - much higher than that of element ratios (Tab. 1). We, therefore, assume that systematic errors due to incorrect background correction disturb our data. This assumption is supported by the observation that most of the variation comes from comparing different sections of the same specimen, even if placed on the same grid, whereas analyses in different sensilla on the same section are highly consistent.

Testing the performance of background correction by film stacks

The aim of the Hall procedure is to correct any X-ray spectrum for mass thickness variation by dividing the characteristic peak counts of a given element by the continuum radiation generated by the same volume of the specimen. Thus, the peak-to-continuum ratio is independent of mass thickness in specimens of the same composition. As already pointed out by Hall (1971, see also Roomans, 1980), this will be only true if all other continuum contributions, such as extraneous X-ray counts from the microscope column, have been subtracted, so that the continuum radiation considered for quantification comes entirely from the specimen. The Link Quantem FLS program provides facilities to correct for extraneous background from the holder and from the grid by measuring the characteristic element counts of holder (e.g. Cu) and grid (e.g. Ni) elements, and assuming a linear relationship between extraneous continuum radiation from holder and grid and their respective characteristic counts. The conversion factors are obtained by calibrating the instrument with an empty holder and an empty grid, respectively (Gupta and Hall, 1982). It is obvious that the demands for correction of the extraneous background are most exacting for ultrathin sections and with these again for those compartments which have the smallest dry mass content.

In our freeze-dried cryosections very often the mass thickness of the specimen was of the same order of magnitude as that of the support film. Moreover, freeze-drying caused considerable mass thickness variation in extracellular compartments, such as sensillum lymph. In order to check the performance of our analytical set-up for background correction we searched for a suitable test specimen with known variations of mass thickness. Perforating a normal filmed grid usually causes the film to fold back thus producing a double film and, sometimes, even a triple layer. Local thickness variation of the film is probably insignificant, so that the mass thickness of the stacks should follow the ratio of 1: 2: 3; moreover, the "in-hole" counts can be measured as well.

Fig. 3 gives an example of such an experiment. Several features are immediately evident: Firstly, a considerable fraction of the total background counts is not created in the specimen, but comes from extraneous sources, in particular the grid. Secondly, a sub-

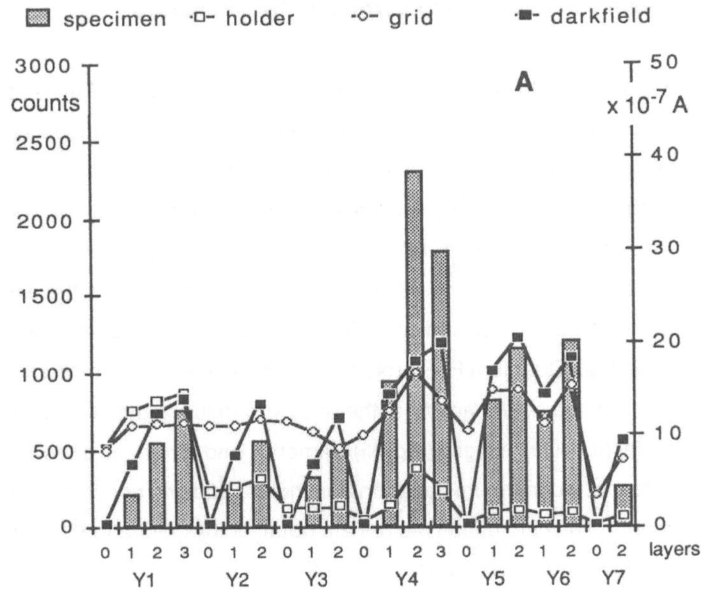

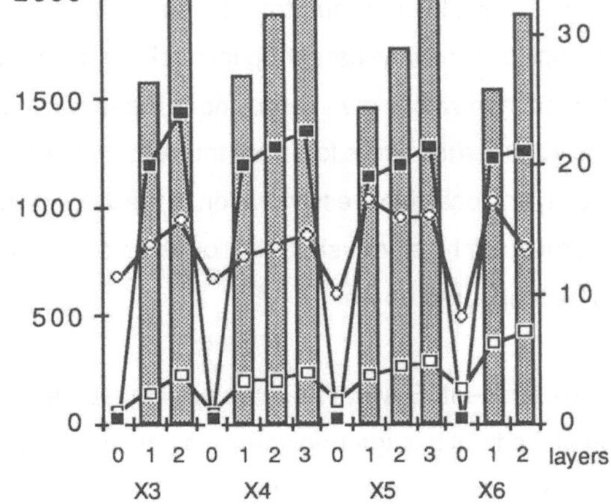

Fig. 3. Continuum radiation as measured in holes, single film, double film, and triple film layers and on different grid squares of a 100-mesh nickel grid. Measurements were done in two rows of grid squares, one (**A**) parallel to the detector axis (vertical row Y1 - Y7) and one (**B**) at right angles (horizontal row X3 - X6), starting with Y1 most distant from the detector. Analyses within a given grid square are from areas close to each other (0, 1, 2, 3 film layers), but the position of the measuring areas relative to the grid bars change from one grid square to the next. The net background integral of the specimen (= film) between 4.2 and 6.2 keV is shown by the bars. It was determined by the Quantem FLS program by subtracting from the total background a holder and grid correction (white symbols) as calculated from the Cu and Ni element counts, respectively. The darkfield intensity as measured by the current on the annular darkfield detector of the STEM is indicated by the black squares (right ordinate). Note that a triple layer was not encountered in every grid square, and that the in-hole spectrum of Y6 was lost, as well as the single film spectrum of Y7. For further explanation see text.

stantial part of this extraneous background is not even dependent on mass thickness variation in the specimen, as demonstrated by the high proportion of in-hole counts and the small increase of holder and grid correction with increasing specimen thickness (=number of film layers). Since the Quantem FLS program is based on the assumption of negligible in-hole counts, it does not provide facilities to correct for such background and a systematic error is introduced into the concentration calculation. In-hole counts may either come from uncollimated electrons or from electrons backscattered below the specimen plane (Nicholson et al., 1982; Warner et al., 1985). The postulated background ratio of 1: 2: 3 for the background created in single, double, and triple layers of the specimen is satisfactorily fullfilled only in the first three squares Y1 - Y3. In square Y4, however, all values except for the in-hole counts are substantially increased; moreover, the background ratios are discontinuous (1:2.4:1.9, instead of 1:2:3). Also in the squares Y5 and Y6 the specimen background ratios are highly different from the postulated 1: 2 (1: 1.4 and 1: 1.6). In Y7, finally, all values are extremely low. This is the grid position next to the detector, and shielding of X-rays by the rim of the holder might have occurred. Also in the horizontal row (X3-X6) of grid squares the measured specimen background would follow the postulated 1: 2: 3 ratio only, if a constant number of counts would be subtracted from each group. Interestingly this number is close to that of the measured in-hole counts.

The signal of the darkfield detector (Fig. 3, line connecting black squares) closely follows the ratio of the specimen background data in the "good" squares of Y1-Y3. Moreover, also in the other squares the discrepancies from the postulated ratios are of the same sign and order of magnitude as those of the net specimen background counts.

Some of the observed variation of the specimen background obviously depends on the position of the analysis area on the grid, e.g. distance from grid bars or holder, as can be seen from the spatial dependency of the grid and holder correction (see also Roomans, 1988). In Y4, however, this interpretation is unlikely because of the sudden increase of *all* measured data. It is more likely that temporal changes in the specimen current have occured (in our instrument the emission of the field emission gun is constantly recorded and controlled, but the specimen current cannot be measured). Thus, not only spatial, but also temporal influences during analysis may considerably influence the background measurements. The latter are even more dangerous, since they cannot be predicted, and obviously may easily escape the attention of the operator during analysis.

Spray-frozen standard droplets

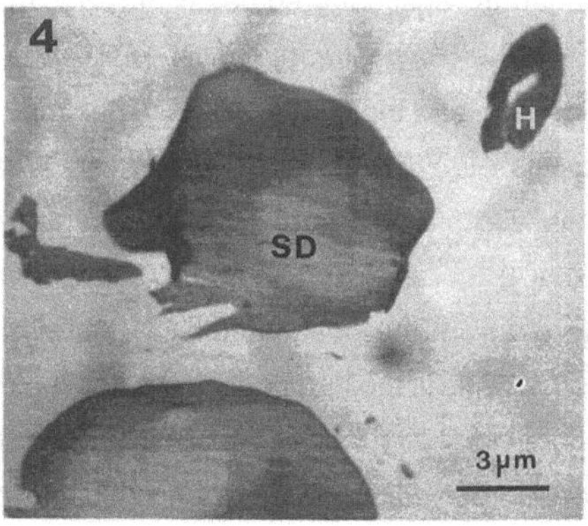

Fig. 4. Freeze-dried cryosections of spray-frozen standard droplets (SD) and of a sensory hair (H).

As a consequence of the foregoing, frequent control of instrument performance appears to be of paramount importance. This should be done with every section and at frequent intervals. It usually is, however, only possible, if every section contains an internal standard of known composition, a pre-condition which rarely is given. The application of a peripheral standard to the specimen before cryofixation has been successfully exploited by several groups (Dörge et al., 1978; Gupta and Hall, 1979; Rick et al., 1982; for a general evaluation see also Hall and Gupta, 1982), but not in every case proof is available that no artificial changes are evoked in the physiology of the specimen by applying the standard to the living tissue.

The ideal solution to this problem would be a standard that could be added to the specimens *after* cryofixation. The heptane cryo-embedding method provides a facility to do this. Frozen standard solutions can be co-embedded with the cryofixed specimen and later can be sectioned together in the same block. We took advantage of the fast cooling rates of the spray-freezing method to obtain optimally cryofixed standard droplets of ~10 µm diameter. Salt solutions of known concentration were mixed with dextran to give a dry mass fraction of 20%. These were spray-frozen by closely following the original procedure of Bachmann and Schmitt (1971), except for using heptane as a glue instead of butyl benzene. In a second step the cryofixed antennae were mounted at -89° C in a droplet of heptane full of frozen standard droplets.

The sectioning properties of the standard droplets are not yet satisfactory and the yield of useful sections is as small as with sections of sensory hairs (Fig. 4). Possibly, this yield could be substantially improved by a better method of section transfer from knife to grid, by which very small sections and section fragments are no longer lost. Although still not perfect, this method of co-embedding frozen standard droplets with specimens was already very useful in checking the performance of the analytical set-up during analysis. It also allows peripheral standard calibration in addition to the continuum method, and thereby provides for another means of calculating the water content in the specimen. Precondition, of course, is that the thickness of the standard section is the same as that of the tissue section and that non-uniform shrinking upon freeze-drying does not occur (Hall and Gupta, 1982).

Conclusions

This paper should be taken as one more caveat to observe quantitative X-ray microanalytical data with caution. Although most of the pitfalls are well known, there is still a need for objective and independent criteria to assess wether in a given set of analyses the quantitative data are reliable or not. Admittedly the given examples are more or less anecdotical and come from the worse end. Without doubt thicker sections will reduce the observed errors considerably, but if we want to take advantage of the high spatial resolution achievable in ultrathin cryosections, the demands for the analytical performance will be extremely severe.

REFERENCES

Bachmann L, Schmitt WW (1971) Improved cryofixation applicable to freeze etching. Proc natl Acad Sci USA 68:2149-2152

De Kramer JJ, Hemberger J (1987) The neurobiology of pheromone reception. In: Prestwich GD, Blomquist GJ (eds) Pheromone biochemistry. Academic Press, New York, pp 433-472

Dörge A, Rick R, Gehring K, Thurau K (1978) Preparation of freeze-dried cryosections for quantitative X-ray microanalysis of electrolytes in biological soft tissues. Pflügers Arch 373:85-97

Gupta BL, Hall TA (1979) Quantitative electron probe X-ray microanalysis of electrolyte elements within epithelial tissue compartments. Fed Proc 38:144-153

Gupta BL Hall TA (1982) Electron probe X-ray microanalysis. In: Baker PF (ed) Techniques in cellular physiology, pt2, P128.Elsevier/North-Holland Biomed Press, Amsterdam, pp 1-52

Hall TA (1971) The microprobe assay of chemical elements. In: Oster G (ed) Physical techniques in biological research, 2nd edn., vol IA. Academic Press, New York, pp 157-275

Hall TA, Gupta BL (1982) Quantification for the X-ray microanalysis of cryosections. J Microsc (Oxford) 126:333-345

Kaissling KE, Thorson J (1980) Insect olfactory sensilla: structural, chemical and electrical aspects of the functional organization. In: Sattelle DB, Hall LM, Hildebrand JG (eds) Receptors for neurotransmitters, hormones and pheromones in insects. Elsevier/North-Holland Biomed Press, Amsterdam, pp 261-282

Nicholson WAP, Gray CC, Chapman JN, Robertson BW (1982) Optimizing thin film X-ray spectra for quantitative analysis. J Microsc (Oxford) 125:25-40

Rick R, Dörge A, Thurau K (1982) Quantitative analysis of electrolytes in frozen dried sections. J Microsc (Oxford) 125:239-247

Roomans GM (1980) Problems in quantitative X-ray microanalysis of biological specimens. Scanning Electron Microsc 1980/II:309-320

Roomans GM (1988) The correction for extraneous background in quantitative X-ray microanalysis of biological thin sections: some practical aspects. Scanning Microsc 2:311-317

Schneider D (1957) Elektrophysiologische Untersuchungen von Chemo- und Mechanorezeptoren der Antenne des Seidenspinners *Bombyx mori* L. Z vergl Physiol 40:8-41

Schneider D (1984) Insect olfaction - our research endeavor. In: Dawson WW, Enoch JM (eds) Foundations of sensory science. Springer, Berlin-Heidelberg-New York, pp 381-418

Steinbrecht RA (1984) Cryofixation and follow-up techniques in biological electron microscopy - outline and demonstration. Zeiss MEM 3:9-17

Steinbrecht RA (1987) Functional morphology of pheromone-sensitive sensilla. In: Prestwich GD, Blomquist GJ (eds) Pheromone biochemistry. Academic Press, New York, pp 353-384

Steinbrecht RA, Zierold K (1984) A cryoembedding method for cutting ultrathin cryosections from small frozen specimens. J Microsc (Oxford) 136:69-75

Steinbrecht RA, Zierold K (1987) The electrolyte distribution in insect olfactory sensilla as revealed by X-ray microanalysis. Ann New York Acad Sci 510:638-641

Warner RR, Myers MC, Taylor DA (1985) Inaccuracies with the Hall technique due to continuum variation in the analytical microscope. J Microsc (Oxford) 138:43-52

Zierold K (1988) X-ray microanalysis of freeze-dried and frozen-hydrated cryosections. J Electron Microsc Techn 9:65-82

Zierold K, Steinbrecht RA (1987) Cryofixation of diffusible elements in cells and tissues for electron probe microanalysis. In: Steinbrecht RA, Zierold K (eds) Cryotechniques in biological electron microscopy. Springer, Berlin-Heidelberg-New York, pp 272-282

Acknowledgement We thank Miss S. Dongard for skilful technical assisitance

Conference Discussion

A.J. Spencer: How are the spray-frozen droplets incorporated into the heptane?

R.A. Steinbrecht: We used the same procedure and apparatus as described in the original method of Bachmann and Schmitt (1971) with the only exception that heptane was used instead of butyl-benzene to glue them together. These lumps were stored under LN_2 and later used in the cryochamber of the cryoultramicrotome for cryoembedding the specimen.

R.D. Leapman: What energy range of continuum was used to obtain quantitative results? There is an advantage in using a lower energy window for the continuum (around Al K α). This is much less sensitive to spurious radiation because of soft X-rays generated on the underside of the grid being absorbed, whereas harder X-rays are transmitted through the grid and pass into the detector (see work of C.E. Fiori).

R.A. Steinbrecht: The continuum was measured in the range of 4.2 - 6.2 keV. We chose this range because around 1.5 keV the energy window which can be used without interference by characteristic peaks (Mg at 1.25 keV, Si at 1.74 keV) is rather small. This would cause an increase of the error in the measured continuum count value. However, your arguments for this low energy range in continuum measurements are conceivable as well.

Progress in Electron Energy Loss Spectroscopic Imaging and Analysing Biological Specimens with a Field Emission Scanning Transmission Electron Microscope

C. Colliex, C. Jeanguillaume, C. Mory, and M. Tencé,
Laboratoire de Physique des Solides associé au CNRS,
Université Paris-Sud, Bâtiment 510, 91405 Orsay, France.

INTRODUCTION

A major problem raised in electron microscopy imaging and analytical studies on biological sections is the conflict between high dose requirements for getting useful signal/noise values on the different channels of information and low dose requirements for beam damage reduction. Among various solutions, our approach consists in developing a strategy which optimize the signal extraction associated to the scattering process suffered by any primary electron. The basic design of the Scanning Transmission Electron Microscope (STEM) is well adapted because it is possible, for each probe position on the specimen, to filter and simultaneously record different fractions of the transmitted beam, discriminated in scattering angle with a set of variable apertures and in energy loss with an EELS spectrometer. Electron energy loss spectroscopy constitutes then a basic component of the microscope as pointed out as early as 20 years ago by Crewe (1966). It can be used fruitfully either as a contrast-enhancement device or as a real analytical tool for chemical microanalysis, depending on the recorded fraction of the EELS spectrum. In a previous publication (Colliex et al. (1984)), we had described unconventional modes for STEM imaging of biological structures. The present paper intends to update it, covering two specific aspects: novel imaging methods and analytical microscopy. The basic methodology has not changed noticeably but the concomitant progress in detection design and in data read-out and processing have enlarged the field of applications and improved the limits of detection by at least an order of magnitude.

INSTRUMENTAL DEVELOPMENTS

The basic features of a STEM instrument such as built by Vacuum Generators are : i) a field emission gun ; ii) a versatile illumination system made of two condenser lenses and one high excitation probe forming lens ; iii) a detection system made of an annular dark field detector which collects all electrons scattered to large angles, and of a spectrometer followed by a bright field detector which analyses sequentially the energy loss spectrum in the transmitted beam. With this microscope, variable primary 100 kV electron probes are available on the specimen, with minimum diameter down to $\simeq$ 0.5 nm and current up to a few 10^{-10} A. Around the Orsay STEM, recent efforts have been made on multidetector configuration and computer assistance.

*1) **Multidetector configuration***

In order to enhance the detection efficiency on the bright field channel at the exit of the EELS spectrometer, a magnetic lens (with double gap and rotation free) has been introduced to provide a magnified image of the spectrum on a multichannel detector. Recent realizations have shown that parallel EELS acquisition either with a SIT tube (Shuman and Kruit (1985)) or on a parallel diode array (Krivanek et al. (1987)) or a two dimensional charge coupled device (Strauss et al. (1987)) constitutes a great improvement compared to the traditional serial detection for many applications in spectroscopy, more specially concerning analysis with core edges. These detectors are made of a number of simultaneous detection cells (typically 1024 in a linear array). But they are not very well adapted for high spectral intensities in the unscattered or low energy loss beams. In an intermediate step, we have inserted in the final spectrum plane a double detector, made of two similar scintillator-photomultiplier chains such as shown in fig. 1a. With a given setting of the magnet excitation, one can record on bf 1 the unscattered beam and on bf 2 the inelastic beam. If one adds the adf signal and the primary current measured on the illumination aperture, four signals are then simultaneously available, the interest of this combination will be emphasized further on.

*2) **Computer assistance**.*

The STEM configuration offers unique capabilities, such as recording :

.) single images ;

.) duplicated images, which reproduce the measurement without changing any instrumental parameter and which are useful to evaluate global image properties such as resolution and signal to noise ratio by cross correlation techniques (Frank (1980)) ;

.) assemblies of simultaneous images through the four channels described above ;

.) sequences of images, with one parameter such as the energy loss value being modified between them, in order to provide characteristic elemental images (Jeanguillaume et al. (1978), Bonnet et al. (1988)) ;

.) any adequate combination of the previous images, such as the sequences of duplicated pairs of images, which are used to estimate the spatial resolution in elemental images (Mory et al. (1988)).

For an efficient handling of all these possibilities, a specific interface has been realized between the electron microscope and the computer (Tencé et al. (1987)). A scan generator includes three independent ramp generators for the three commands : spectrometer, line and frame, which can be addressed following any desired sequence. The input unit acquires, analyses and stores the informations provided by the different detectors. This system has proved to be very powerful in the different unconventional imaging and analysing modes developed in the laboratory.

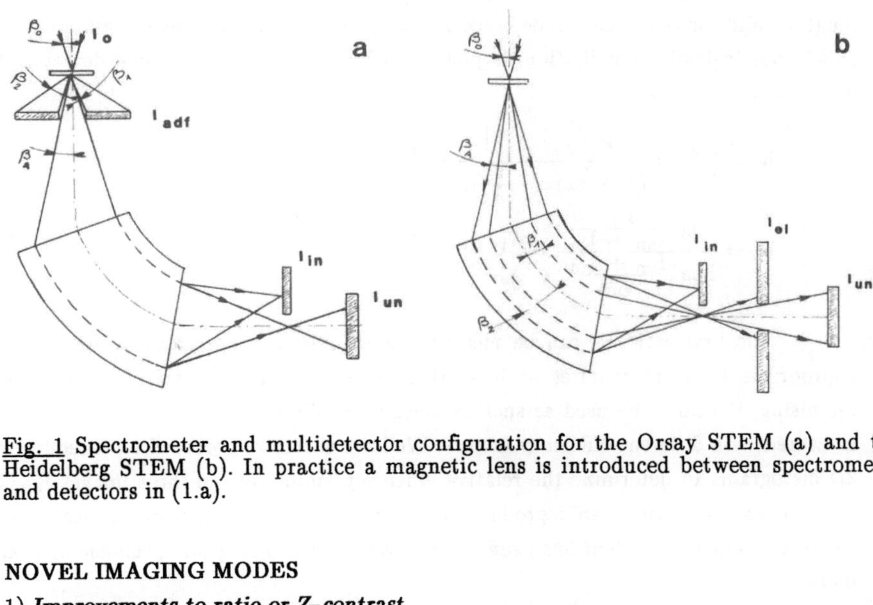

Fig. 1 Spectrometer and multidetector configuration for the Orsay STEM (a) and the Heidelberg STEM (b). In practice a magnetic lens is introduced between spectrometer and detectors in (1.a).

NOVEL IMAGING MODES

1) *Improvements to ratio or Z-contrast*.

The so-called atomic number contrast corresponds to the ratio between the annular dark field I_{adf} and the inelastic component I_{in} of the bright field. It has been initially used by Crewe and coll. for imaging single high Z atoms deposited on thin supporting carbon layers (Isaacson et al. (1980)). A few years ago, this method has been applied to the visualization of ultrastructures in thin unstained biological sections (Carlemalm et al. (1984)). However it was shown that this contrast is not really thickness independent (Egerton (1982), Reichelt and Engel (1984), Colliex et al. (1984)) : only in the case of very thin samples the ratio contrast depends more sensitively on the chemical composition than on thickness variations. It is therefore very important to improve the technique in order to extend its range of applicability to thicker sections. Egerton (1982) proposes different routes. Using his notations with λ_i and λ_e refering respectively to the inelastic and the elastic mean free paths, t being the local thickness, and I_{un} the unscattered signal :

$$I_{un} = I_0 \exp(-\frac{t}{\lambda_e}).\exp(-\frac{t}{\lambda_i})$$

$$I_{in} = I_0 \exp(-\frac{t}{\lambda_e}).[1 - \exp(-\frac{t}{\lambda_i})]$$

$$I_{adf} = I_0 [1 - \exp(-\frac{t}{\lambda_e})]$$

These equations, which neglect the detailed angular efficiency factors, are roughly valid for the situation depicted in fig. (1a). All the high angle scattering, with or without inelastic contribution, is picked up by the annular dark field detector. We introduce :

$$R_1 = \frac{I_{adf}}{I_{in}} = \frac{[\exp(\frac{t}{\lambda_e}) - 1]}{[1 - \exp(-\frac{t}{\lambda_i})]} \rightarrow \frac{\lambda_i}{\lambda_e} \text{ for } t \rightarrow 0$$

$$R_2 = \frac{\text{Log}\,\frac{I_0}{I_{un} + I_{in}}}{\text{Log}\,\frac{I_{un} + I_{in}}{I_{un}}} = \frac{\lambda_i}{\lambda_e}$$

The first ratio R_1 can be measured with only two detectors, while the more appropriate R_2 term requires at least three detectors. Though this formula is very promising, it cannot be used as such because it involves additions of terms which are measured with different efficiency factors. We have therefore proposed a procedure with 2D histograms to determine the relative efficiency factors of the three images I_{un}, I_{in}, I_{adf}, so that their sum can reproduce the I_0 map (Jeanguillaume and Tencé (1987)). The success of the method has been demonstrated on a biological specimen with knife marks.

Another solution has been suggested by Egerton (1982) and brought to reality only recently by Haider (1988). It consists in separating the unscattered and elastically scattered electrons after the spectrometer as shown in fig. 1b. I_{el} is then a filtered annular dark field image. However, the angular acceptance of the spectrometer is very large and the inelastic detector picks up all inelastic events, whether or not they are associated to large angle elastic scattering. The corresponding formulae are then :

$$I_{un} = I_0 \exp(-\frac{t}{\lambda_e}).\exp(-\frac{t}{\lambda_i})$$

$$I_{el} = I_0 \exp\left(-\frac{t}{\lambda_i}\right).\left[1 - \exp\left(-\frac{t}{\lambda_e}\right)\right]$$

$$I_{in} = I_0 \left[1 - \exp\left(-\frac{t}{\lambda_i}\right)\right]$$

Similarly to above, two types of ratios can be measured, involving three signals :

$$R_3 = \frac{\mathrm{Log}\,\dfrac{I_{un} + I_{el}}{I_{un}}}{\mathrm{Log}\,\dfrac{I_0}{I_{un} + I_{el}}} = \frac{\lambda_i}{\lambda_e} = R_2$$

(already proposed by Egerton (1982)), or two signals, the filtered dark field image divided by the inelastic one, as suggested by Haider (1988) :

$$R_4 = \frac{I_{el}}{I_{in}} = \frac{\left[1 - \exp\left(\frac{t}{\lambda_e}\right)\right.}{\exp\left(\frac{t}{\lambda_i}\right) - 1} \rightarrow \frac{\exp\left(-\frac{t}{\lambda_e}\right.}{\exp\left(\frac{t}{\lambda_i}\right)}.R_1$$

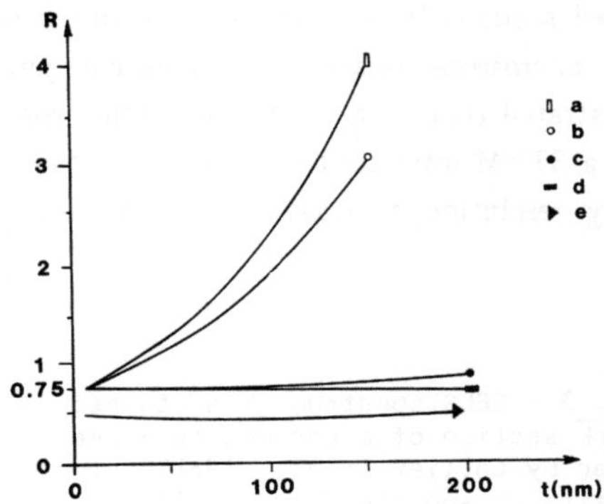

Fig. 2 - Calculation of the different ratio contrasts introduced in table I, as a function of thickness. Specimen is amorphous carbon. Incident electron energy 100 keV (courtesy of M. Haider, unpublished 1988).

Table I

	R_1 (a)	R_1 (b)	R_2 (c)	R_3 (d)	R_4 (e)
β_0	1 mrad	8.3 mrad	1 mrad	1 mrad	7.5 mrad
β_1	13 mrad	13 mrad	10 mrad	3 mrad	13 mrad
β_2	130 mrad	130 mrad	130 mrad	130 mrad	100 mrad
β_A	5 mrad	10 mrad	10 mrad	130 mrad	100 mrad
ΔE	5-80 eV	5-80 eV	5-150 eV	5-150 eV	5-50 eV

which has the same $\frac{\lambda_i}{\lambda_e}$ limit in the very thin specimen case. Within this approximate description with two parameters, λ_i and λ_e, R_4 would exhibit a reverse behaviour with respect to R_1, i.e. it would decrease with thickness instead of increase. Haider (private communication) has improved this model by taking into account the angular and energy dependence on the various measured beams. Some of his results are gathered in fig. 2 for a carbon foil. The parameters listed in Table I have been considered. The curve (b) would correspond roughly to the experimental conditions used in Orsay. While the rather flat behaviour obtained for (e) with the Heidelberg machine is due to the reduced width of the energy window which limits the weight of the inelastic term for strongly increasing thickness. Nevertheless, the complete combination of three signals seems far superior, if cleared from eventual artifacts associated to unsufficient knowledge of the different efficiency factors. It would be possible to apply it for compositional determination even in rather low dose conditions because of the involved short mean free path, if compared with core-loss excitations.

2) ***Imaging of thick sections***.

In order to investigate the three dimensional organization of cell biologic organelles, thick embedded and stained sections (t $\gtrsim$ 0.5 μm) are required but they present special problem to electron microscope imaging. In a recent publication (Colliex et al. (1988), we have demonstrated that energy filtered imaging constitues a useful solution. When performed with a STEM interfaced to a good data collection set up, quantitative behaviours of intensity, resolution, contrast, SNR can be measured.

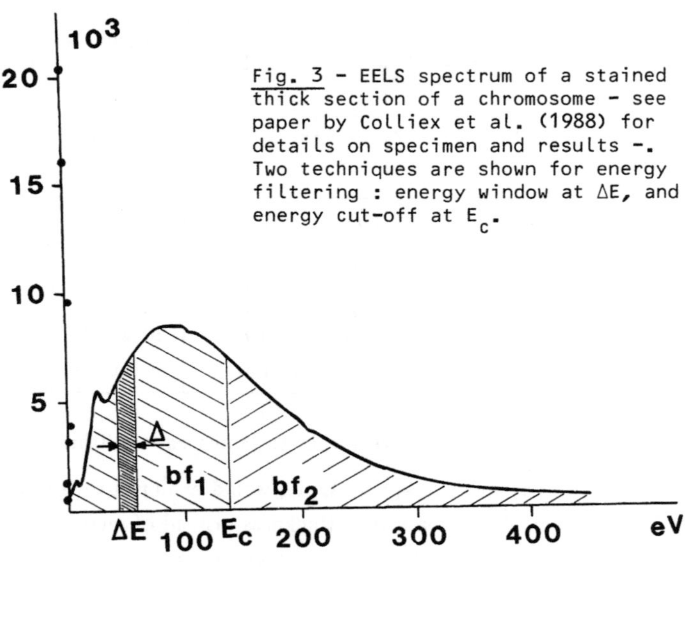

Fig. 3 - EELS spectrum of a stained thick section of a chromosome - see paper by Colliex et al. (1988) for details on specimen and results -. Two techniques are shown for energy filtering : energy window at ΔE, and energy cut-off at E_c.

Figure 3 explains the principle of the method. This is an experimental spectrum of a 0.4 μm thick section of a chromosome. Apart the narrow elastic peak culminating at about $20{\times}10^3$ counts, the curve shows a dominant inelastic contribution at about 100 eV due to multiple inelastic scattering. In the first series of experiments, an energy window of width $\Delta\ eV$ was used to record filtered images and the position ΔE of this slit was scanned along the spectrum, in order to determine optimum conditions for the visualization of different cellular components within the nuclei of salivary gland visualization. This method can also be used with a filtering CTEM, such as the Zeiss 902, as demonstrated by Bauer et al. (1987).

In order to fully take benefit of the absence of chromatic aberration in the STEM, Pearce-Percy and Cowley (1976) had suggested an alternative approach, also schematized in fig. 3 : a cut-off energy loss E_c is defined which separates the spectrum in two halves (the low energy loss and the high energy loss domains). With the system operated in Orsay (fig. 1a), it is possible to collect simultaneously the two signals : the spectrometer is set so that bf_1 is collected by the I_{un} and bf_2 by the I_{in} detector. Figure 4 gathers three micrographs recorded simultaneously with these detection channels. We are presently investigating the behaviour of any of these images as a function of E_c but is is clear from now that an order of magnitude in SNR can be gained with respect to the previous energy window technique. Many possibilities, involving linear combinations or correlations of these images, can be proposed to obtain more accurate maps of λ_i/λ_e, or of any of these quantities separately.

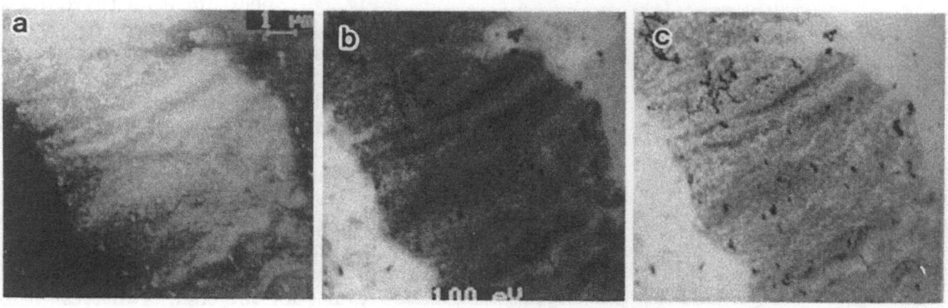

Fig. 4 Three micrographs recorded simultaneously on channels I_{adf}, I_{un} (or bf_1) and I_{in} (or bf_2) from the same area of a chromosome thick section ($t \simeq 0.4$ μm). E_c is 100 eV.

DEVELOPMENTS IN ANALYTICAL MICROSCOPY

On this subject, a comprehensive and complete paper has recently been published by Leapman and Ornberg (1988) and the review written by Jeanguillaume (1987) presents an extensive compilation of all life science applications performed with EELS on sections or on isolated molecules and organelles.

1) ***Advances in spectroscopy***.

It concerns the recording and processing of spectra acquired with a probe fixed or rastered over a reduced area, encompassing the feature of interest (molecule, lysosome, membrane...). The elemental or chemical information is contained in the intensity of the various core edges (N, O, F, P...) which can be identified on the continuously decreasing background and it constitutes the major use for EELS up to now. But the molecular signature, which lies in fine structures on core edges or in the low loss region, should also be readily accessible with the advent of parallel EELS systems, as a consequence of the reduced radiation damage for an equivalent signal to noise.

Over the past recent years, improvements have mostly concerned the quantitative measurement of biologically important elements and the detection of small concentrations. To illustrate one specific aspect, fig. 5 and table II present typical data recorded with sequential detection on a biological section. The basic quantitation scheme follows the standard Egerton's technique (1978).

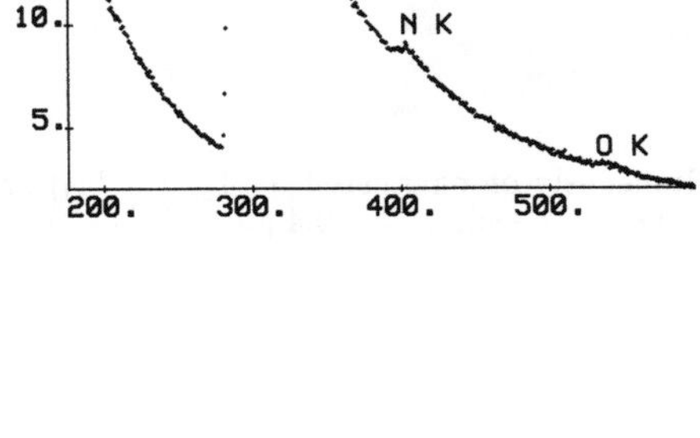

Fig. 5 - Spectrum from thyroid gland in epon (sample provided by E. Larras), showing Cl-L_{23} (due to embedding material), C-K, N-K and O-K edges. Quantitative analysis is performed for one signal below and one above the dominent carbon edge (table II). With conventional EELS techniques (sequential spectrum acquisition and background stripping following a power law model), detectable concentrations lie in the range 0.1 to 1 % with better results for the edge of interest below the matrix one than above.

	Signal	Background	SNR	C (% at)	Detection limits for first and second kind risks = 0.05	Atomic cross-section σ (25 mrad, 80 eV) in cm^2
C	1 160 000	211 000	890	92.3 ± 0.1		8.6×10^{-21}
Cl	131 300	447 000	31.2	2.3 ± 0.1	0.26 %	39.5×10^{-21}
N	33 000	508 395	9.4	5.4 ± 0.2	2.1 %	4.2×10^{-21}

Table II

The used software, developed by Trebbia (1988) also contains a statistical analysis of the random fluctuations. It provides a good estimate of the variance of the signal, so that the SNR of the measurement can be estimated and extrapolated to the detection limits, taking into account both the first and second kind risk of errors (for missing an element present or finding an element absent). Of course this constitutes only one aspect of the experimental accuracy, other factors such as the knowledge of the cross-sections and the occurrence of selective mass loss have also to be considered.

Other approaches for quantitation are presently being introduced in EELS (Shuman and Somlyo (1987), Leapman and Swyt (1988)). One simulates the experimental spectrum with a family of reference edges recorded on standards ; the finite thickness influence is taken into account by convoluting with the low energy loss spectrum and concentrations are determined as the weighting factors of the different edges in the best least squares fit. When applying these techniques to parallel recorded spectra, Shuman and Somlyo (1987) have demonstrated an approximate detection limit of 1 mmol Ca/kg drywt in an organix matrix (≃ 12 ppm) with 100 s counting time and doses of a few 10^9 e/nm^2 into a 100 nm probe. This result constitutes a great jump forward because it makes accessible to EELS quantification, the biological relevant range of 1 to 100 mmol of Ca/kg drywt.

2) ***Advances in mapping***.

With the development of computer controlled data acquisition systems, elemental mapping from sequences of energy filtered images has become more quantitative. Softwares have been implemented, which permit an accurate EELS background determination and subsequent core edge measurement at each pixel in the image, together with a statistical analysis of the variance of the background and of the signal (Bonnet et al. (1988)). Consequently one can nowadays display 2D maps of any of the involved parameters (Signal, Background, SNR...) and the way is opened towards real quantification of the number of atoms/pixels.

The fundamental limitation in this case is the constraint of a reduced counting time/pixel if one is interested by the spatial distribution of the information over a reasonable image size (i.e. 256^2 pixels). When used in high spatial resolution conditions, mostly limited by the 0.5 nm probe size, Mory and Colliex (1988) have demonstrated that one could evaluate numbers of uranium atoms per pixel, with a standard deviation of one or two units. This constitutes a great potentiality in the context of heavy metal labeling of active sites on molecules, as already demonstrated by Monson et al. (1987) for 12-tungsten and 11-gold stable clusters.

Finally, a new concept in EELS digital acquisition and processing - the spectrum image - is presently emerging (Jeanguillaume et Colliex (1988)). It is a 3D matrix of numbers, two axis correspond to the (x,y) position on the specimen and the third one is associated to a complete EELS spectrum provided by a parallel EELS device. Though they are of very great size, these spectrum-images can be handled with normal computing technology, in order to extract from various sections or projections, a chemical information more directly connected to the morphological structure.

Acknowledgements :

Thanks are due to P. Ballongue for his assistance in the design and construction of the multidetector configuration, to P. Trebbia and N. Bonnet for their contribution in the development of useful softwares, to E. Delain, A. Olins, E. Larras and C. Quintana for their collaboration on various biological problems and to M. Haider for his fruitful correspondence and communication of data on novel imaging modes before publication.

References

Bauer R, Hezel U, Kurz D (1987), High resolution imaging of thick biological specimens with an image electron energy loss spectrometer, Optik 77: 171-174.

Bonnet N, Colliex C, Mory C, Tencé M (1988), Developments in processing image sequences for elemental mapping, In : Image and signal processing in electron microscopy, Scanning Microscopy, Supplement 2: 351-364.

Carlemalm E, Colliex C, Kellenberger E (1984), Contrast formation in electron microscopy of biological material, In : PW Hawkes (ed) : Advances in Electronics and Electron Physics, Academic Press 63: 269-334.

Colliex C, Jeanguillaume C, Mory C (1984), Unconventional modes for STEM imaging of biological structures, J. Ultrastr. Res. 88: 177-206.

Colliex C, Mory C, Olins AL, Olins DE, Tencé M (1988), Energy filtered STEM imaging of thick biological sections, to be published in J. of Microscopy, 151.

Crewe AV (1966), Scanning electron microscopes : is high resolution possible ?, Science 154: 729-738.

Egerton RF (1978), Formulae for light element microanalysis by EELS, Ultramicroscopy 3: 243-251.

Egerton RF (1982), Thickness dependence of the STEM ratio image, Ultramicroscopy 10: 297-300.

Frank J (1980), The role of correlation techniques in computer image processing, In: PW Hawkes (ed), Computer processing of electron microscope images, Topics in Current Physics, Springer-Verlag 13: 187-222.

Haider M (1988), Filtered dark field and pure Z-contrast: two novel imaging modes in a STEM, In: C Colliex and MS Isaacson (eds): Proc. NSF/CNRS Workshop on Electron beam induced spectroscopies with high spatial resolution, Aussois: 333-342.

Isaacson M, Utlaut M, Kopf D (1980), Analog computer processing of STEM images, In: P Hawkes (ed), Computer processing of electron microscope images, Topics in Current Physics, Springer Verlag, 13: 257-283.

Jeanguillaume C, Colliex C, Trebbia P (1978), About the use of EELS for chemical mapping of thin foils with high spatial resolution, Ultramicroscopy 3: 237-242.

Jeanguillaume C (1987), Electron energy loss spectroscopy and biology, Scanning Microscopy 1: 437-450.

Jeanguillaume C, Tencé M (1987), How to obtain a thickness independent image in a STEM, Ultramicroscopy 23: 63-76.

Jeanguillaume C, Colliex C (1988), Spectrum-image: the next step in EELS digital acquisition and processing, In: C Colliex and MS Isaacson (eds): Proc. NSF/CNRS Workshop on Electron beam induced spectroscopies with high spatial resolution, Aussois: 349-357.

Krivanek OL, Ahn CC, Keeney RB (1987), Parallel detection electron spectrometer using quadrupole lenses, Ultramicroscopy 22: 103-116.

Leapman RD, Ornberg RL (1988), Quantitative electron energy loss spectroscopy in biology, Ultramicroscopy 24: 251-268.

Leapman RD, Swyt CR (1988), Unravelling of overlapping core edges in EELS, In: C Colliex and MS Isaacson (eds): Proc. NSF/CNRS Workshop on Electron beam induced spectroscopies with high spatial resolution, Aussois: 322-324.

Monson KL, Wall JS, Hainfeld JF (1987), Visibility and stability of a 12-tungsten atom complex in the STEM, Ultramicroscopy 21:147-156.

Mory C, Colliex C (1988), Elemental analysis at the single atom detection level by processing sequences of energy filtered images, In: C Colliex and MS Isaacson (eds): Proc. NSF/CNRS Workshop on Electron beam induced spectroscopies with high spatial resolution, Aussois: 472-481.

Mory C, Bonnet N, Colliex C, Kohl H, Tencé M (1988), Evaluation and optimization of the performance of elastic and inelastic STEM imaging by correlation analysis, In: Image and signal processing in electron microscopy, Scanning Microscopy Supplement 2: 329-342.

Pearce-Percy HT, Cowley JM (1976), On the use of energy filtering to increase the contrast of STEM images of thick biological materials, Optik 44: 273-288.

Reichelt R, Engel A (1984), Monte Carlo calculations of elastic and inelastic electron scattering in biological and plastic materials, Ultramicroscopy 13: 279-294.

Shuman H, Kruit P (1985), Quantitative data processing of parallel recorded electron energy loss spectra with low signal to background, Rev. Sci. Instrum. 56: 231-239.

Shuman H, Somlyo AP (1987), Electron energy loss analysis of near trace element concentrations of calcium, Ultramicroscopy 21: 23-32.

Strauss MG, Naday I, Sherman IS, Zaluzec NJ (1988), CCD-based parallel detection system for electron energy loss spectroscopy and imaging, Ultramicroscopy 22: 117-123.

Tencé M, Bonnet N, Colliex C and Mory C (1987), EELS elemental mapping: progress in quantitation and in sensitivity, In: DC Joy (ed), Analytical Electron Microscopy San Francisco Press: 295-298.

Trebbia P (1988), Unbiased method for signal estimation in EELS. Concentrations measurements and detection limits in quantitative microanalysis: methods and programs, Ultramicroscopy 24: 399-408.

Conference Discussion

K. Zierold: Can the methods you describe for imaging of thick sections be used for quantitative EELS or elemental mapping of approximately 100 nm thick cryosections? After freeze-drying, these sections exhibit strong variations in local thickness!

C. Colliex: The methods of energy loss filtering, either with a slit or with a cut-off method, give strong contrast for thick sections. But this contrast is rather dependent on t/λ, consequently it is not well adapted to discriminate various components in the cryosections from strong variations in local thickness. On a theoretical basis, I would rather recommend the multisignal strategy with ratio signals which should extend the range of applications of these methods to specimens of 100 to 200 nm (see Fig. 2). I must confess that the experimental confirmation is to be realized.

K. Zierold: To which extent is it possible to measure K by EELS in the presence of much C, as it is the usual situation in biological specimens?

C. Colliex: Detection and measurement of low potassium concentrations constitute actually a very difficult task for EELS techniques, as the potassium L_{23} threshold at $\simeq$ 294 eV lies just on the maximum of the excitation of the carbon K edge. Consequently, the standard background extrapolation cannot be used in this situation. Hopefully, the potassium L edge is characterized by a sharp edge with white lines and a great fraction of the signal is contained within a narrow band of about 10 eV. A more useful approach has recently been developed by Shuman and Somlyo (Ultramicroscopy 21, 23 (1987)) and by Leapman and Swyt (to be published in Ultramicroscopy (1988)). It consists in a multiple-least-squares fitting with previously recorded reference spectra. In their paper "Separation of overlapping core edges in EELS spectra by multiple-least-squares fitting", Leapman and Swyt deduce from their experimental results that it is thus possible to quantitate about 0.2 at% of potassium in a thin carbon matrix. This performance is much better than previously demonstrated with other methods. However, the real quantitation of low levels of potassium in organic matrix has not yet been tested and these authors recognize that slight variations in carbon K edge shape might degrade the quality of the fit.

Application of Parallel-Detection Electron Energy Loss Spectroscopy in Biology

Richard Leapman
National Institutes of Health
Bethesda
Maryland 20892
U.S.A.

Introduction

For many years it has been appreciated that electron energy loss spectroscopy (EELS) is the most sensitive microanalytical technique for detecting the biologically ubiquitous light elements, carbon, nitrogen and oxygen (Isaacson and Johnson, 1975). In particular, elemental ratios such as nitrogen-to-sulfur or nitrogen-to-phosphorus can provide information about relative concentrations of proteins (containing sulfur-rich amino-acids) and nucleotides (containing phosphate groups) in subcellular organelles (Leapman and Ornberg, 1988). With conventional EELS instrumentation it has been relatively easy for example to measure nitrogen present typically at levels of 5 to 10 atomic percent but phosphorus and sulfur occurring at levels below about 1 to 3 atomic percent are often only weakly visible in the core loss spectrum. Other interesting elements like calcium and iron have not generally been detectable at physiological concentrations using EELS until the recent advent of highly efficient parallel detection systems. Now it is becoming apparent that EELS may in fact compete with and even surpass the performance of energy-dispersive x-ray spectroscopy (EDXS) for such elements (Shuman and Somlyo, 1987). The aim of this paper is to compare directly the sensitivity of EELS and EDXS, to illustrate new methods of spectral processing, and to consider the state-of-the-art and future prospects for biological applications.

Parallel-Detection Spectrometers

Optimized detection limits for electron energy loss spectroscopy require simultaneous detection of the inelastically scattered electrons:

previously only low efficiency serial detection was available. Several types of parallel recording systems have recently been developed (e.g., Shuman, 1981; Johnson et al., 1981; Egerton, 1984; Krivanek et al., 1987) each having certain advantages. Important design characteristics to have emerged include: (i) use of post-spectrometer quadrupole lenses to magnify the dispersion so that resolution is not limited by channel separation (typically 25 μm); (ii) indirect exposure of the array (e.g. using radiation-resistent YAG scintillators); (iii) channel-to-channel gain variations not greater than about 1%; (iv) large dynamic range (>10^3); (v) detector quantum efficiency (DQE) close to 1; (vi) fast read out times (<20 ms) and low read-out noise.

We have employed a Gatan (model 666) parallel electron energy loss (PEELS) system based on a 1024-channel photodiode array (Krivanek et al., 1987) and combined with a Gatan (model 607) magnetic sector electron energy loss spectrometer. The spectrometer is attached to a Hitachi H700H analytical electron microscope with a scanning transmission (STEM) accessory and a Tracor Northern Microtrace energy dispersive x-ray detector. Electron and x-ray spectra can be acquired simultaneously by means of a Tracor Northern TN5500 microanalysis system which controls the integration time, number of read-outs, and the voltage offset to the isolated drift tube of the Gatan PEELS.

Spectral Acquisition and Processing

In order to detect very weak core edges it is essential to remove channel-to-channel gain variations from the PEELS spectrum. Several methods are available for doing this as described in detail by Shuman and Kruit (1985). One approach, that we adopt here, is to record the second difference spectrum which is equivalent to applying a top-hat digital filter, thereby enhancing sharp features while suppressing the slowly-varying background. The filtering operation is achieved by acquiring three spectra with $-\Delta$, $+\Delta$ and zero voltages applied to the spectrometer drift-tube. The sum of the negative- and positive-offset spectra is subtracted from twice the zero-offset spectrum which removes channel-to-channel gain at the same time. The magnitude of the offset Δ must be carefully matched with the width of the feature to be detected since it determines the level of smoothing in the result.

The top-hat filter is similar to that used for quantitating x-ray spectra but, whereas in EDXS spectra peak widths stay quite constant, energy resolution can vary in EELS spectra making quantitation more difficult.

A multiple least squares (MLS) fitting algorithm can be used, as for analysis of EDXS spectra, to quantitate the second difference (or unfiltered) PEELS spectra. However, for energy loss spectra the method must take into account plural inelastic scattering as described by Shuman and Somlyo (1987), and by Leapman and Swyt (1988). Reference spectra must first be obtained from pure compounds, e.g. salts deposited on carbon films. Plural scattering reference spectra are generated by convoluting the carbon low-loss (plasmon) spectrum with the 'single scattering' core edge spectra. Using this approach Shuman and Somlyo (1987) have demonstrated that it is possible to quantitate 25 ppm of calcium in a carbon matrix.

Comparison of EELS with EDXS : Predictions for Fe and Ca:

It is useful to estimate a relative figure of merit between parallel EELS and a well optimized EDXS system. The ratio of signals for the two spectroscopies is given by,

$$S_{EELS}/S_{EDXS} = (\sigma_{EELS}/\sigma_{EDXS}).(f_{EELS}/f_{EDXS}).(\eta_{EELS}/\eta_{EDXS}).(A_{EELS}/\omega_{EDXS}) \qquad (1)$$

In this expression σ is the appropriate total ionization cross section of which only fraction f can be utilized for microanalysis. The geometric collection efficiency for the signal is η. In EELS the signal is further attenuated by a thickness-dependent factor A due to plural scattering whereas in EDXS the signal is reduced by the x-ray fluorescence yield ω. If the signal-to-background ratio is Q (defined by the peak area/background within the full width at half-maximum) then we can write an approximate expression for the ratios of signal-to-noise as,

$$(S/N)_{EELS}/(S/N)_{EDXS} = [(S_{EELS}/S_{EDXS}).(Q_{EELS}/Q_{EDXS})]^{1/2} \qquad (2)$$

In EELS iron and calcium are best detected by the L_{23} edge but at low concentrations (signal/background<<0.01) only the 'white line' resonance intensity close to threshold is useful: further away from the edge uncertainties in the background become important due to extended fine structure or to plural inelastic scattering. Experimentally we have determined that f_{EELS}=0.044 for iron and f_{EELS}=0.083 for calcium. For EDXS analysis we choose the K_{α} line for which the relative transition probability is f_{EDXS}=0.9. In our microscope the collection angle for EDXS is 0.1 steradians and we assume the DQE for EELS is near unity. Then using total cross sections calculated from Powell (1985) for a beam energy of 100 keV and using fluorescence yields from Heinrich (1981) we find that S_{EELS}/S_{EDXS} is 580 for iron and 2760 for calcium. The ratio of signal/backgrounds in EELS and EDXS can be obtained from simultaneously acquired spectra. Using measured values of Q_{EDXS}/Q_{EELS} = 500 (Fe) and Q_{EDXS}/Q_{EELS} = 300 (Ca), we arrive at $(S/N)_{EELS}/(S/N)_{EDXS}$ = 1.1 and 3.0 for Fe and Ca respectively. Therefore we expect that the two techniques should be comparable for iron but that EELS should be more sensitive for calcium, provided samples are thin enough for plural scattering not to dominate the spectrum.

Experimental Results

A. Iron in Cryosectioned Erythrocytes

The unfiltered PEELS spectrum from a red blood cell (Fig. 1a) in a freeze-dried cryosection of rat liver is shown in Fig. 1b. Data was collected in the STEM mode at 100 keV beam energy by rastering over a 100 nm square region of a cell having a thickness of approximately 1.2 inelastic mean free paths, as measured from the low loss spectrum (Leapman et al., 1984). The spectrum was acquired with 200 read-outs, with an integration time of 200 ms (total time 40 s) and with a probe current of approximately 2 nA. Since one count corresponds to about 20 high energy electrons there are 3×10^6 to 3×10^7 electrons per 0.5 eV channel in the spectrum, but very few features are discernible in the raw data apart from the strong carbon K edge and a small shoulder at the nitrogen K edge. The second difference spectrum in Fig. 1c was acquired under similar conditions but now the N K, O K and Fe L_{23} edges are easily visible at 400 eV, 531 eV and 710 eV respectively even though the Fe signal/background is only about 0.01. In Fig. 1d the MLS fitting procedure is applied to the Fe L_{23} edge and we find that the atomic fraction Fe/C = $6.7 \pm 0.6 \times 10^{-4}$. The EELS result for iron can be

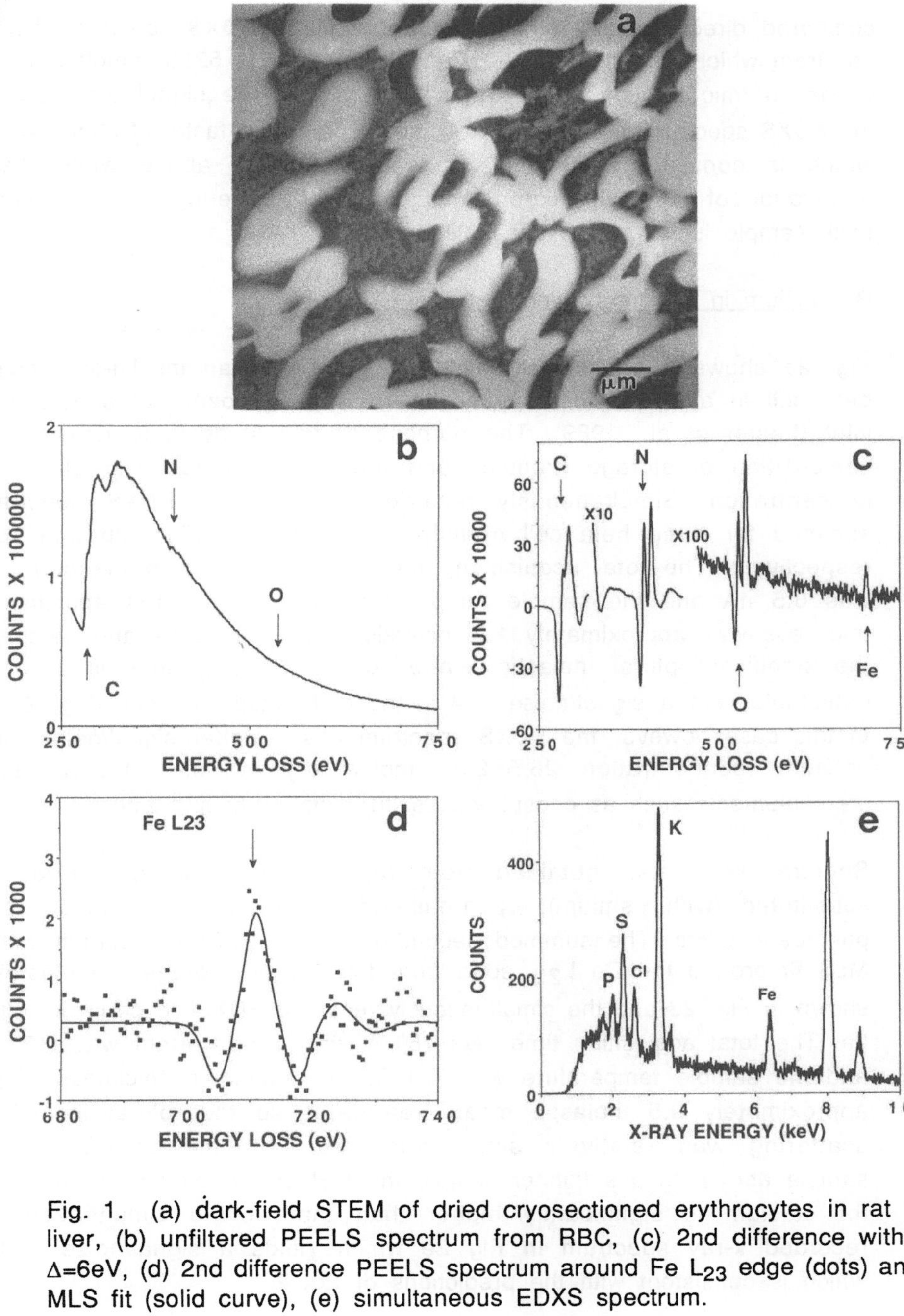

Fig. 1. (a) dark-field STEM of dried cryosectioned erythrocytes in rat liver, (b) unfiltered PEELS spectrum from RBC, (c) 2nd difference with Δ=6eV, (d) 2nd difference PEELS spectrum around Fe L_{23} edge (dots) and MLS fit (solid curve), (e) simultaneous EDXS spectrum.

compared directly with a simultaneously acquired EDXS spectrum (Fig. 1e) from which we find that the Fe concentration is 52±3 mmol/kg dry wt. (or atomic fraction ~ $6.3 \pm 0.3 \times 10^{-4}$). So the Fe signal-to-noise in the EDXS spectrum is higher than in the PEELS by a factor of about two, which is consistent with our calculated estimate above when the attenuation of the EELS core signal by plural scattering in this rather thick sample is taken into account.

B. Calcium in Pancreatic Islets of Langerhans

Fig. 2a shows a bright-field STEM micrograph of an insulin-secreting beta cell in a freeze-dried cryosection of rapidly frozen rat pancreatic islet (Foster et al., 1989). The purpose of this study is to determine composition of storage granules and how this changes with glucose concentration. Simultaneously recorded PEELS and EDXS spectra summed for three beta cell granules are shown in Figs. 2b and 2c respectively. The total acquisition time was 360 s, the probe current was 0.5 nA and the sample temperature was -70 C. The specimen thickness was approximately 1.2 inelastic mean-free paths and despite the significant plural inelastic scattering the Ca L_{23} edge is clearly detectable with a signal/noise ~ 4 in the MLS fitted spectrum (Fig. 2b). In this case however the EDXS spectrum has a better signal/noise for calcium (concentration 26.5±2.3 mmol/kg dry wt.) and also reveals other elements such as phosphorus, sulfur, potassium and zinc.

Spectra were also obtained from rapidly-frozen, acetone freeze-substituted (with osmium), epon-embedded, and uranium stained rat pancreatic islets. The summed second-difference PEELS spectrum with MLS fit around the Ca L_{23} edge from five insulin storage granules is shown in Fig. 2d and the simultaneously recorded EDXS spectrum in Fig. 2e. The total acquisition time was 150 s, the probe current was 0.5nA and the sample temperature was -70 C. The specimen thickness was approximately 0.5 inelastic mean free paths so that plural inelastic scattering was relatively small compared with the cryosectioned sample above. In this thinner sample the PEELS signal/noise ratio of 7 for calcium is significantly higher than that in the simultaneously recorded x-ray spectrum in Fig 2e which yields a signal/noise ~ 3, which is consistent with the predictions of Eq. 2.

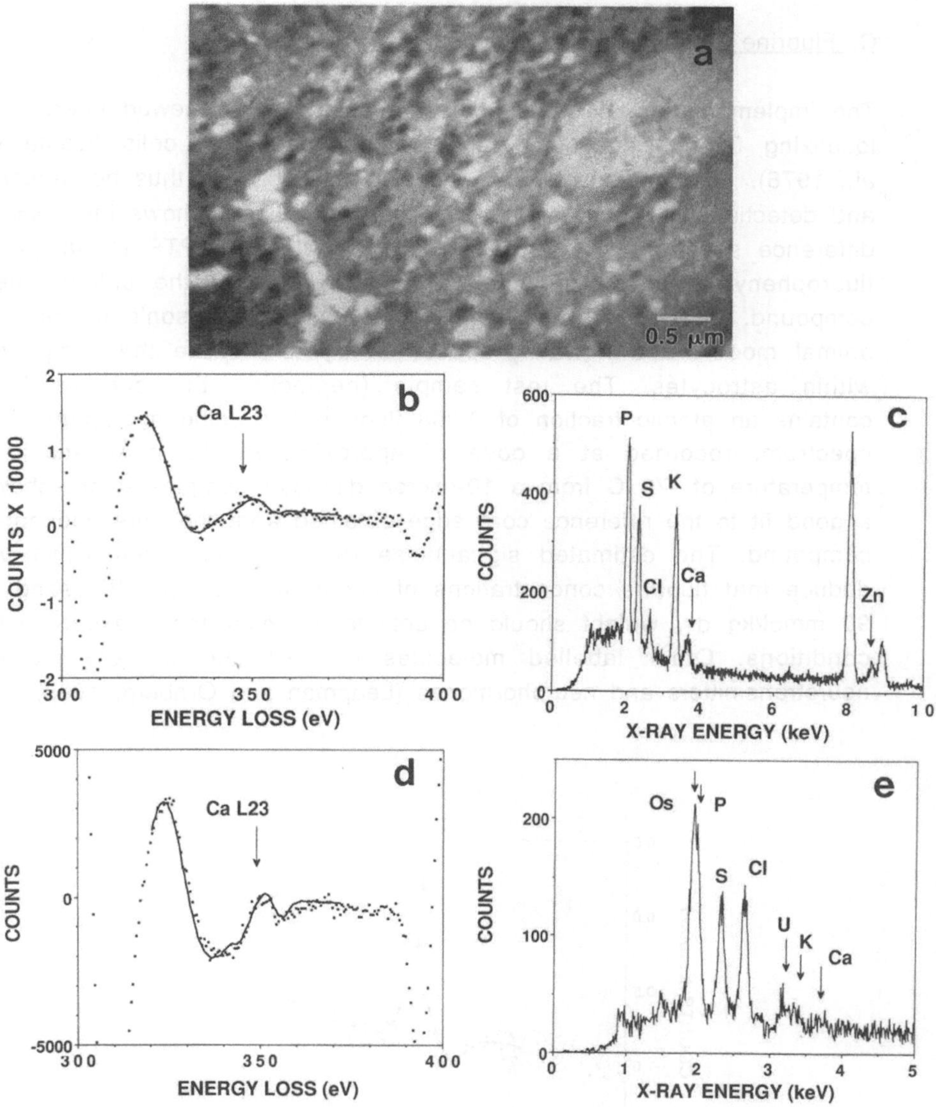

Fig. 2. (a) bright-field STEM of dried cryosectioned rat pancreatic islet, (b) 2nd difference PEELS (Δ=6eV) of three beta cell granules in cryosectioned islet (dots) and MLS fit to C K and Ca L_{23} edges (solid curve), (c) simultaneous EDXS spectrum from cryosection, (d) PEELS data from five beta cell granules in thinner freeze-substituted plastic section of rat islet (sample provided by Drs. A.F. Boyne and R.W. Dudek), (e) simultaneous EDXS spectrum from plastic section.

C. Fluorine Detectability in Labelled Molecules

The implementation of parallel-detection EELS has renewed interest in localizing fluorine-labelled organic compounds within cells (Costa et al., 1978). Fluorine loss due to radiation damage can thus be reduced and detection limits correspondingly improved. Fig. 3 shows the second difference spectrum from a sample of 10% fluoro- MPTP (1-methyl 4-fluorophenyl tetrahydro-pyridine) mixed with 90% of the unfluorinated compound. MPTP may be useful for studying Parkinson's disease in animal models and it was proposed to try to localize the compound within astrocytes. The test sample (neglecting the support film) contains an atomic fraction of 0.008 fluorine (excluding hydrogen). The spectrum, recorded at a dose of approximately 10^3 e/Å^2 and at a temperature of -70 C from a 10-micron diameter sample region, shows a good fit to the reference core edge obtained from the pure fluorinated compound. The estimated signal/noise ratio is 16 : 1 from which we deduce that fluorine concentrations of about one part per thousand or 90 mmol/kg dry weight should be detectable under these experimental conditions. Other labelled molecules of particular interest include neurotransmitters and neurohormones (Leapman and Ornberg, 1988).

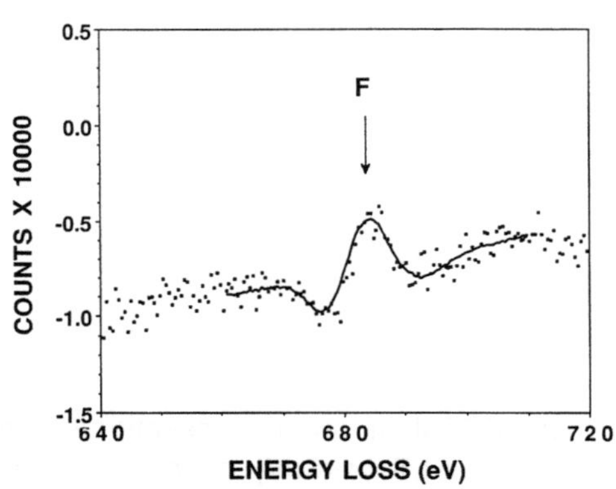

Fig. 3. 2nd difference PEELS (Δ=6eV) from 10% fluoro-MPTP (dots) and MLS fit to F K edge (sample provided by Drs. S.P. Markey and A. Marini).

Prospects for Future Work and Conclusions

An important recent development is the acquisition of elemental maps in the STEM by collecting entire core edge spectra while digitally rastering the electron probe over an array of points in the specimen. Previously elemental mapping was only possible with serial detection (Jeanguillaume et al., 1983; Leapman et al, 1984; Colliex et al., 1985). Parallel detection has now brought a significant improvement (Shuman et al., 1986; Hunt, 1988; Krivanek, 1988) and with inexpensive computer technology it is now possible to record entire spectra at each pixel. This has the advantage of allowing *a posteriori* data processing incorporating procedures such as (i) correction for spectral energy drift during the image acquisition, (ii) detection of elements that were originally not suspected to be present, and (iii) using more sophisticated spectral processing algorithms that cannot be run in real-time. The main limitation for biological applications is still the relatively long acquisition time required to gather chemical images serially in STEM, even with new detectors that can gather the spectrum in parallel. This situation can be contrasted with the energy-filtering TEM where large images can be recorded rapidly in parallel, but the spectral information must be collected serially (Ottensmeyer, 1984). Nevertheless, PEELS mapping has two important advantages over the energy-selected TEM technique: it can be performed simultaneously with x-ray mapping; and radiation damage can be minimized.

To summarize, parallel detection EELS provides the most sensitive means for measuring low atomic number elements at high spatial resolution. In addition, the technique competes favorably with EDXS for elements such as calcium, iron, sulfur and phosphorus. Some thinner cryosections can be analyzed successfully at 100 keV beam energy but results could generally be improved by operating at 200 or 300 keV to reduce attenuation of the core edge signal by plural scattering.

Colliex C, Jeanguillaume C, Mory C (1984) Unconventional modes for STEM imaging of biological structures. J Ultrastruct Res 88:177-206

Costa JL, Joy DC, Maher DM, Kirk KL, Hui SW (1978) Fluorinated molecule as a tracer: difluoroserotonin in human platelets mapped by

EELS. Science 200:537-539
Egerton RF (1984) Parallel recording systems for electron energy loss spectroscopy (EELS). J Electron Micros Tech 1:37-52
Foster MC, Leapman RD, Li MX, Ornberg RL, Atwater I (1989) (in press) Composition of secretory granules by EPMA of cryosectioned rat pancreatic islets. Biophys J (abstract)
Heinrich KFJ (1981) Electron beam x-ray microanalysis, Van Nostrand Reinhold, New York
Hunt J (1988) (to be published) PEELS imaging using an IBM PC AT
Isaacson M, Johnson D (1975) The microanalysis of light elements using transmitted energy loss electrons. Ultramicroscopy 1:33-52
Jeanguillaume C, Tence M, Trebbia P, Colliex C (1983) EELS chemical mapping of low-Z elements in biological sections. In: Johari O (ed) Scanning Electron Microscopy. SEM, AMF O'Hare, IL, Vol. 2, p.745-756
Johnson DE, Monson KL, Csillag S, Stern EA (1981) An approach to parallel-detection EELS. In: Geiss RH (ed) Analytical Electron Microscopy. San Francisco Press, San Francisco, CA, p.205-209
Krivanek OL, Ahn CC, Keeney RB (1987) Parallel detection electron spectrometer using quadrupole lenses. Ultramicroscopy 22:103-116
Krivanek OL (1988) (private communication)
Leapman RD, Fiori CE, Gorlen KE, Gibson CC, Swyt CR (1984) Combined elemental and structural imaging in a computer-controlled AEM. Ultramicroscopy 12:281-292
Leapman RD, Ornberg RL (1988) Quantitative electron energy loss spectroscopy in biology. Ultramicroscopy 24:251-268
Leapman RD, Swyt CR (1988) Separation of overlapping core edges in electron energy loss spectra by multiple least squares fitting. Ultramicroscopy (in press)
Leapman RD, Fiori CE, Swyt CR (1984) Mass thickness determination by electron energy loss for quantitative x-ray microanalysis in biology. J Microsc 133:239-253
Ottensmeyer FP (1984) Electron spectroscopic imaging: parallel energy filtering and microanalysis in the fixed-beam electron microscope. J Ultrastruct Res 88:121-134
Powell CJ (1985) Inner shell cross sections In: Mark TD, Dunn GH (eds) Electron Impact Ionization. Springer-Verlag, NY, p.199-231
Shuman H, Kruit P(1985) Quantitative data processing of parallel recorded electron energy loss spectra with low signal to background. Rev Sci Instrum 56:231-239
Shuman H (1981) Parallel recording of electron energy loss spectra. Ultramicroscopy 6:163-168
Shuman H, Somlyo AP (1987) Electron energy loss analysis of near-trace element concentrations of calcium. Ultramicroscopy 21:23-32
Shuman H, Chang CF, Buhle EL, Somlyo AP (1986) EELS: quantitation and imaging. Ann NY Acad Sci 483:295-310

Conference Discussion

B.L. Gupta: You have shown us data for C, N, O, K. Ca etc. How about Na? We all believed that a great advantage of EELS would be in the softer energy bands where the efficiency of EDX is rather poor.

R.D. Leapman: We have not yet examined this question in detail. I suspect, however, that the advantage of EELS (high collection efficiency) versus X-ray spectroscopy (low fluorescence yield) for sodium detection may not be as great as one might think. Firstly, it is necessary to use the K shell excitation in EELS so there is no advantage in the ionization cross section for EELS relative to that for X-ray spectroscopy. Secondly, for low Na concentrations (tens of mM/kg) the EELS signal/background ratio is extremely small: under these conditions an important parameter for EELS detectability is the fraction of the total cross section that is near the Na/K edge and unfortunately, this is rather low. Thirdly, it is now possible to use ultra-thin window detectors to gain additional sensitivity for sodium detecton by X-ray spectroscopy.

J. Shelburne: In your comparison of EELS and EDX, you assume a single EDX detector. Particularly in scanning electron microscopes (where there is usually plenty of room), why not have multiple EDX detectors?

R.D. Leapman: It should certainly be possible to fit more than one EDX detector on the microscope column. High resolution microanalysis of ultrathin crysections is generally performed in the STEM mode in an analytical transmission electron microscope. In this type of instrument there is some practical limitation for the number of detectors that can be fitted around the sample because of the presence of the objective pole-piece, specimen holder, specimen drive mechanism, objective aperture, anticontaminator, etc. I agree that more effort should be made to increase the collection solid angle of the X-ray detector, which at present is rarely greater that 1.5 %.

P. Ingram: In evaluating the efficacy of EDX vs EELS for the detection and measurement of calcium, how do you justify using an osmium-fixed and stained sample where the presence of the heavy atoms could absorb the Ca X-rays?

R.D. Leapman: The osmium-fixed section was very thin (about 40 nm) and was only lightly stained, so it is unlikely that any significant absorption of calcium Kα X-rays would occur. Clearly, this type of sample is unsuitable for quantitative cell physiology and is merely intended as an experimental comparison between EELS and EDX for calcium detectability in a thin sample.

R. Rick: Using an EDX system, probably the biggest problem in cellular Ca measurements is the overlap with the K-ß peak of potassium. Is there a similar overlap in EELS using your curve fitting approach?

R.D. Leapman: There is no significant problem with overlap of the Ca L_{23} edge with the K L_{23} edge in EELS because the structures are well separated in energy. However, there may be overlap of the very large carbon K edge, particularly when plural scattering is significant (in thicker samples) or when extended fine structure oscillations cannot be neglected. Although these effects may be very small (a fraction of 1 % of the spectral intensity), the calcium L_{23} edge may be even smaller at physiological calcium concentrations.

N. Roos: You state that EELS might have a three times better detectability for some elements compared to X-ray microanalysis. However, this seems to be set off by the need to have to use extremely thin specimens and very high electron doses in order to acquire a spectrum. Could you comment on this?

R.D. Leapman: The electron doses used to compare experimentally the sensitivity of EELS and EDX were exactly the same: the spectra were recorded simultaneously. It is true, however, that relatively thin samples are required to optimize sensitivity in EELS. The specimen thickness needs to be less than the inelastic mean free path (about 60 nm of carbon at 100 keV beam energy). At thicknesses greater than this, the signal/noise ratio deteriorates rapidly. At 100 keV, beam energy we have found that typical dried cryosections are often approximately

1.5 to 2 times the optimum thickness. We use a folding copper grid with formvar/carbon on both sides to sandwich the sections so a significant improvement could be made by minimizing the support film thickness. Another possibility is to increase the microscope beam voltage to 200 or 300 keV. Then I believe that EELS would generally have higher sensitivity than EDXS for calcium in thin dried cryosections. But such a comparison may not be fair if EDX works well at 100 keV. Perhaps EELS will be most suited to the highest resolution studies on the thinnest and best quality cryosections that can be prepared.

Resin Based Standards for Biological Energy Dispersive X-Ray and Electron Energy Loss Microanalysis

D.C.Joy, C.S.Joy and D.A.Armstrong
EM Facility
University of Tennessee
Knoxville
TN 37996-0810
USA

Introduction

Elemental standards, or calibration specimens, can perform several important functions in biological microanalysis. Firstly for the quantification of EDS data from ultrathin biological section using Halls' method (1971) they provide a means of determining the weight factor for a given element. Secondly they provide an objective and reproducible way of testing the performance of the analytical system in such areas as its absolute sensitivity and its freedom from system induced artifacts (Williams and Steel 1987). Thirdly they provide a library of peaks and edges for use in the iterative reconvolution quantification techniques developed for application to cases where severe peak or edge overlaps occur in a biological system (Leapman and Swyt 1987), for use in statistical tests to determine the validity of an edge in a noisy spectrum (Joy 1982), and for use as comparison spectra for such cases as those where the valence state of an element in an unknown is to be determined. In order to meet these needs in our own laboratory we have been developing a family of calibration specimens containing elements of biological interest at concentrations from 10 to in excess of 100 mmol/Kg.

Properties Required of a Calibration Specimen and Standard

A calibration specimen for microanalysis must be well defined chemically, must be homogeneous and free from precipitates or extraneous inclusions at the desired spatial scale of a high resolution microanalysis i.e below ten nanometers (Andersen 1967). Finally the behavior of the calibration sample as a function of the electron beam irradiation to which it is subjected must be well documented and any changes in elemental concentration or total mass must be monotonic with dose and sufficiently small to permit accurate correction of these quantities to zero-dose conditions by a linear extrapolation. If, in addition, the calibration sample is to be referred to as a *standard* then its composition must remain homogeneous at a sufficiently bulk scale so as to permit an independent macroscopic assessment of the chemical composition of the material and the referencing of this composition to an agreed and traceable standard medium.

Ideally, because quantitation methods often rely on the use of the bremsstrahlung or energy loss continuum spectrum as a measure of specimen mass-thickness (e.g Hall 1968), the calibration sample should resemble the typical biological section in its chemical composition and physical properties as closely as possible since otherwise large, and poorly understood, corrections may be required (Lauchli 1972). An ideal standard would have the same matrix and composition as the specimens of interest. A material based on the addition of a suitable compound to an epoxy or acrylic resin produces a calibration sample which is a close match for the analysis of tissue embedded in resin and so several systems of this type have been described (e.g Ingram and Hogben 1968; Kriz et al 1972; Spurr 1974, 1975; Jessen et al 1974; Chandler 1976; Roomans and van Gaal 1977; Harvey et al 1980; Colonna et al 1982; Condron and Marshall 1986). The work discussed here builds on these previous examples to try to generate calibration samples which meet the criteria listed above and which, at the same time, avoid some of the problems encountered by earlier workers.

Choice of Materials

The element of interest must be added to the resin in the form of some suitable compound which must, itself, satisfy a variety of conflicting requirements. It must be sufficiently soluble in the resin that the desired concentration of the element can be reached before saturation occurs and leads to precipitation and gross inhomogeneity. Care must also be taken to ensure that the compound is not unevenly distributed in the resin, so producing a gradient of concentration across the width of the block (Condron and Marshall 1986). Because the resultant block must be ultra-microtomed in order to produce sections of a usable thickness it is also important that the addition of the chosen compound does not adversely affect the normal cutting properties of the resin (Harvey et al 1980). Finally the compound should be chosen to meet the goal of minimizing the radiation sensitivity of the resin. This implies the use, where possible, of high molecular weight compounds containing unsaturated or ring structures since these factors are known to promote resistance to both elemental translocation and mass-loss during electron irradiation.

Attempts have been made to use the intrinsic elemental contents of resins, such as Cl and S, for calibration purposes but such materials are clearly special cases. A more general approach is to make the standard by the incorporation of a metal complex or salt with ligands capable of solubilizing metals in organic media. On the evidence of previous investigations (Ornberg and Reece 1981) we have concentrated on the naphthenate and octoate group of ligands. These compounds possess a promising combination of properties for this type of applications. They are available as liquids at room temperature, and are soluble in common embedding resins. This facilitates the incorporation of the element of interest and alleviates the problem of insoluble extraneous substances in the resulting resin block. Naphthenates or octoates can be formed for most metals of interest, and such compounds are readily and inexpensively available, and are likely to remain so, because of their industrial applications. Finally the chemistry of the naphthenates and octoates in resins is well characterized, and consequently their interaction with the resin may be predicted and controlled.

TABLE I

		Modified Spurr's	Temp.	LR White	Temp
K				10 - 100 mmol/kg	RT
Ca		30 - 100 mmol/kg	340K		
Fe		100 - 300 mmol/kg	RT		
Zn	Naphthenates	100 -300 mmol/kg	340K	10 - 100 mmol/kg	RT
Co		100 - 300 mmol/kg	RT	10 -100 mmol/kg	RT
Cu				10 - 100 mmol/kg	RT
K		100 - 600 mmol/kg	340K		
Zn	Octoates	100 -300 mmol/kg	340K		
Ni		100 - 300 mmol/kg	340K		

Naphthenic acids are not, as has sometimes been believed, single compounds but are obtained from crude oils and represent a mixture of cyclopentane carboxylic acid and alkanecarboxylic acids ranging in molecular weight from approximately 100 Daltons to in excess of 1000. Cyclopentane-carboxylic acid is among the lower molecular weight acids present in these mixtures. Naphthenic acids are primarily used in industry as their metal salts, their greatest application being as driers in coatings and resins. They accelerate the change from liquid to solid state by catalyzing the oxidative cross-linking reaction involved. The catalytic activity depends on the metal cation of the soaps and is greater for metals capable of existing in more than one oxidation state and exhibiting greater oxygen transport capability. Octoates are a class of metallic driers prepared from 2-ethyhexanoic acid and may be used if the naphthenate of a desired metal is not available. A list of available commercial metal naphthenates and octoates is given in Table I, and these represent a good source of liquid reagents for the incorporation of metals into organics (Meinstein and Spielman 1968).

Preparation of samples

Samples have been made in both modified Spurr's (Pallaghy's) and LR White resin using a procedure based on that reported by Simon et al (1986). The naphthenates are provided in the form of a solution in mineral spirits containing a specified weight % content of the metal. The excess spirits are removed by evaporation overnight in an oven held at $340^{o}K$. In our experience formation of a "skin" on the surface of the spirits indicates that too much liquid has been removed. The presence of a sediment in the liquid should also be watched for, and removed by ultra-centrifuging if necessary, as this can produce extraneous solids in the final sample. The solution is finally reweighed before mixing it with the appropriate amount of resin which has been prewarmed to $340^{o}K$. The mixture is then stirred on a hot plate, watching carefully for signs of the formation of a cloudy precipitate. In general when a precipitate forms the batch must be discarded although in some cases, e.g the zinc naphthenates, the precipitate can be redissolved by stirring. Otherwise stirring is continued until solution is complete, a process which typically takes one to two hours. The catalytic activity of the metal salt may influence the rate at which the resin sets, for example 100mmol/kg solution of iron naphthenate will harden in only five minutes at 340 ^{o}K so preventing thorough stirring. This problem can be overcome by holding the resin at room temperature although this also limits the solubility of the metal salt. Table I indicates the temperature at which solution to the specified concentration was achieved. The metal appears to influence the hardness of the resin, and for some combinations of metal and concentration a resin too brittle to cut may be produced (e.g Harvey et al 1980). If a preferred concentration falls in the range where this effect is a problem then some compensation might be achieved by changing the amount of plasticizer added to the resin.

Evaluation of Samples

Batches of each of the materials listed in Table I were prepared as discussed above, at concentrations in the range 10 to 200 mmol/kg, in the form of standard sized capsules. When prepared as discussed above the capsules were visually free from sediment or precipitates although occasional bubbles are evident. Sections for observation, nominally 1000Å thick, were cut at room temperature, mounted on copper grids and examined at either room or liquid nitrogen temperature in an Hitachi H-800 200 keV TEM/STEM fitted with a LaB_6 emitter, a KEVEX Quantum ultra-thin window X-ray detector, a GATAN 607 electron energy loss spectrometer and a GATAN 626 cryo-transfer system.

1.Chemical purity and quantitative elemental concentration

EDS spectra from the standards were recorded for periods of 100 seconds, corresponding to a total irradiation dose of $2x10^5$ el/Å^2. The thin window X-ray detector shows peaks from the carbon and oxygen in the resin, in addition to the desired element and some low level system peaks . No peaks which could be associated with the presence of impurities were found for any of the resins in the list down to the sensitivity limit set by the signal to noise ratio of the spectrum. Since, to a good approximation at a fixed beam energy the peak to background ratio of a given peak depends only on the concentration of the element and not on the local thickness of the matrix, this measurement was used to monitor the relative amount of the element of interest incorporated into the resin. On our system at 200keV these peak to background ratios were in the range from 30 to 200. A comparison of this quantity with the nominal concentration of the element in the section as determined from the preparation procedure shows, as for example in the typical case of the calcium data illustrated in figure 1, that the linearity is excellent.

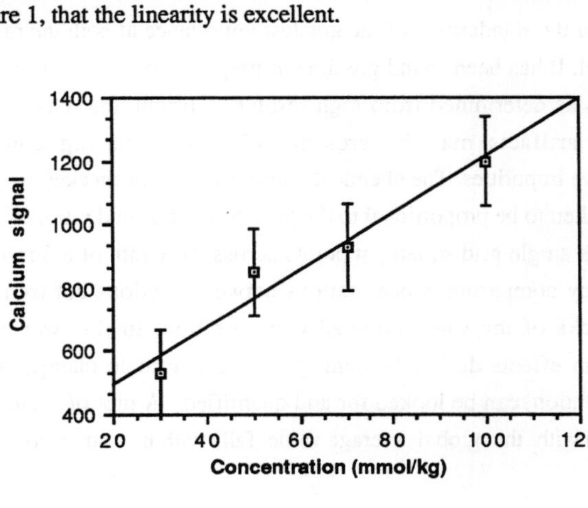

Figure 1:Calcium signal vs nominal concentration

For a dozen measurements on each of four sections of different concentrations the R value was 0.95. Complete blocks (approximately 1 gram in mass) were also analyzed using Atomic Absorption Spectroscopy by a commercial laboratory to provide an independent and macroscopic measure of the elemental concentration. Figure 2 demonstrates, again for the case of calcium, the correlation between this data and the nominal concentration. Although the accuracy with which the nominal concentration is known is not of the highest importance the level of agreement achieved is highly satisfactory indicating that the basic preparation procedure is reliable.

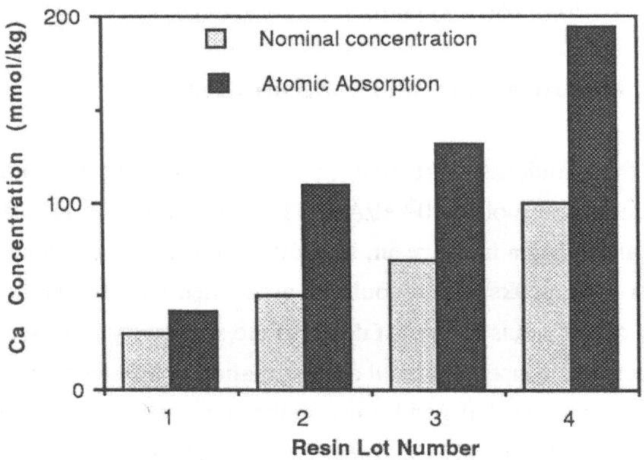

Figure 2: Nominal concentration vs Atomic Absorption data

2.Tests of Homogeneity

The homogeneity of the standards is of the greatest importance at both the microscopic and the macroscopic level. It has been found possible to prepare blocks which are free from precipitates or sediments as determined from high resolution bright and dark field STEM images although such artifacts may be present unless the starting compound is ultra-centrifuged to remove impurities. The chemical homogeneity was checked by measuring the local concentration, taken to be proportional to the peak to background ratio, of the chosen element at points within a single grid square, at points across the width of a 3mm grid for a given section, and also by comparing concentrations between sections cut from different portions of various blocks of the same nominal concentration. In this way both local variations due to random effects during hardening, and larger scale changes caused by incomplete mixing or solution, can be looked for and quantified. A plot of these individual concentrations, compared with the global average value falls within a range consistent with the ±3 standard deviation expected from the average counting statistic. No evidence of systematic deviations from the mean value was observed for any of the samples examined.

3.Radiation behavior

The radiation behavior of the calibration samples was studied by measuring the variation with electron dose of a number of different signals characterizing the specimen. Measures of the mass thickness of the sample were obtained from a calibrated dark-field electron detector, the low loss region of the electron energy loss spectrum, the X-ray continuum between 5 and 6keV, and from the carbon K X-ray peak at 280eV. The composition of the sample was monitored from the characteristic X-ray peak of the element of interest. The beam dose specified represents the irradiated dose incident on the specimen. In the case of the thin section the irradiated and absorbed doses may differ by a factor of 50x or more because a large fraction of the incident beam is transmitted unscattered. Use of the absorbed dose would therefore be preferable since it would ensure that the data was independent of the actual mass-thickness of the section examined, but an online measurement of the absorbed dose was not possible during this sequence of experiments.

The calibration samples all show similar behavior. In their *as cut* state all the sections were found to be excessively prone to mechanical damage by the beam and no reliable measurements were possible. All samples were therefore examined with a coating of about 100Å of carbon on either the beam incidence face, or both faces. At room temperature, and for incident doses up to at least 10^7 el/Å^2 (16000 C/cm^2) the mass-thickness of the chosen area, as determined from the carbon peak or dark field signal, was found to fall exponentially with dose. For a Spurr's resin coated with carbon on one surface only and examined at room temperature the incident dose $D_{1/e}$ required to reduce the mass-thickness to 1/e of its starting value was found to be $3x10^6$ el/Å^2, corresponding to about 100 minutes of continuous beam exposure under the conditions used for spectral acquisition in this paper. The corresponding $D_{1/e}$ for the element incorporated in the resin is typically a factor from 50% less (for potassium) to 30% greater (for calcium) than the value of $D_{1/e}$ for the total mass. This indicates that there is little or no preferential loss of the chosen element relative to the bulk of the matrix. Consequently even over extended periods of measurement on the same area (e.g fifty successive 100 second analyses) the calibration provided by the sample will remain reliable to better than 10%.

If carbon coating is applied to both surfaces of the section then $D_{1/e}$ for the total mass loss is increased by an additional factor of about two, a figure which agrees with that obtained by Rice et al (1987). The mechanism for this improvement is uncertain but may involve the carbon layer acting as a barrier to the loss of the more volatile fractions of the resin components. The $D_{1/e}$ for the elemental concentration is similarly improved by carbon coating on both sides. Measurements were also performed at 100°K using the GATAN cryotransfer holder. To within the accuracy determined by the counting statistics of the experiment the $D_{1/e}$ values were found to be typically 5 to 25% worse than those found for the identical system at room temperature. This result confirms other observations (e.g Simon 1988) that cooling to liquid nitrogen temperatures confers little or no extra protection to the sample. Only in cases where the partial vapor pressure of water in the system is high at room temperature is it likely lowering the temperature, and so freezing out the water vapor and reducing ionization erosion of the sample, will give any significant improvement.

A possible final source of radiation induced problems would be beam enhanced translocation of elements. Such a lateral diffusion would deplete the area under examination of the element of interest at the expense of surrounding regions. Comparative measurements inside an area irradiated for a substantial period (>2500 C/cm^2), and in the immediately surrounding region, showed no evidence of a loss of homogeneity.

Conclusion

Calibration samples based on the solution of naphthenates or octoates in an epoxy resin have been made for seven elements. These samples show excellent homogeneity on both a microscopic and macroscopic scale and can be assayed by a conventional bulk atomic absorption technique to give a precise concentration value. The samples exhibit good stability under the beam at both room and liquid nitrogen temperatures with characteristic dose values $D_{1/e}$ of the order of 3×10^6 el/Å^2.

References

Andersen CA (1967), *An introduction to the Electron Probe Microanalyzer and its application to Biochemistry,* Meth. biochem. Analysis, **15**, 147

Chandler JA (1976), *A method of preparing absolute standards for quantitative calibration and measurement of biological ultrathin specimens in EMMA,* J. Microsc., **106,** 291-302

Colonna KN, Breese SS, Sellers SD and Deck JD, (1982), *An assessment of Ca-Naphthenate-Epon Mixtures as Standards for Calcium Quantification in Aortic Valve*

Leaflets, Proc.40th EMSA Meeting, ed G W Bailey, (San Francisco Press;San Francisco), 390-391

Condron RJ and Marshall AT, (1986), *Inhomogeneity in epoxy resin standards for quantitative biological X-ray microanalysis,* J. Microsc. **143**, 249-255

Hall TA (1968), *The Electron Probe X-ray microanalysis of Thin Specimens,* in "Quantitative Electron Probe Microanalysis", ed KFJ Heinrich, (NBS:Gaithersburg,Md), 269-280

Hall TA (1971) *The Microprobe Assay of Chemical Elements* in "Physical Techniques in Biochemical Research, (Academic Press:NY), **1A**,

Harvey DMR, Flowers TJ, Hall JL and Spurr AR (1980), *The preparation of calibration standards for sodium, potassium and chlorine analyses by analytical electron microscopy,* J. Microsc., **118**, 143-152

Ingram MJ and Hogben AM (1968), *Procedures for the study of biological soft tissue,* Dev.Appl.Spectr., **6**, 43

Jessen H, Peters PD and Hall TA (1974), *Sulphur in different types of keratohyalin,* J.Cell Sci., **15**, 359

Joy DC (1982), *Processing Electron Energy Loss Spectra,* in "Microbeam Analysis-1982". ed KFJ Heinrich, (San Francisco Press:San Francisco), 98-106

Kriz W, Schnermann Y, Hobling HJ, von Rosenstiel AP and Hall TA (!972), *Electron Probe Microanalysis of electrolytes in kidney cells,* in "Recent Advances in Renal Physiology", (Karger:Basel), 162-171

Lauchli A (1972), *Electron Probe Analysis,* in "Microautoradiography and Electron Probe Analysis", ed U Luttge, (Springer:Berlin), 191-200

Leapman RD and Swyt CR (1987), *Removal of Plural Scattering in EELS.* in "Microbeam Analysis-1987". ed RH Geiss, (San Francisco Press:San Francisco), 273-276

Mienstein S and Spielman HR, (1968), *Metallic Soaps,* in "Technology of Paints, Varnishes and Lacquers", ed C R Martens, (Reinhold:New York), 390-421

Ornberg R, Reese T (1981), in "Microprobe Analysis of Biological Systems", ed A V Somlyo, (Academic Press:New York), 213-218

Rice PM, Crozier PA and Egerton RF, (1987), *Control of Mass Loss in analytical electron microscopy,* in "Analytical Electron Microscopy 1987", ed D C Joy, (San Francisco Press:San Francisco), 213-216

Roomans GM and van Gaal HLM (1977), *Organometallic and organometalloid compounds as standards for microanlysis of epoxy resin embedded tissue,* J. Microsc., **109**, 235-240

Simon GT and Heng YM (1986), *Standards for quantification of elements analyzed by energy dispersive spectroscopy and electron spectroscopic imagng*, Proceeding 9th Int.Conf, X-ray Optics and Microanalysis, (Imperial Press:Toronto), 416-419

Simon GT (1988) Personal communication

Spurr AR (1974), *Macrocyclic polyether complexes with alkali elements in epoxy resins as standards for X-ray microanalysis of biological tissues,* in "Microprobe Analysis as Applied to Cells and Tissues", ed TA Hall, PE Echlin and R Kaufmann, (Academic Press:London), 213

Williams DB and Steel EB (1987) *A Standard Cr thin-film Specimen to Measure X-ray Peak to Background Ratio,* in "Analytical Electron Microscopy - 1987", ed D C Joy, (San Francisco Press:San Francisco), 228-229

Conference Discussion

T.A. Hall: You have suggested that the degree of mass-loss in thin specimens depends not only on beam exposure (electrons/area) but on specimen thickness. It seems likely to me (and I hope) that percentage mass loss in thin specimens is independent of thickness. I.e., if you double the specimen thickness, do you get twice as much energy absorbed, and probably twice as many bonds broken and twice as much mass lost, but still the same mass lost/initial mass?

D.C. Joy: I suggested that the appropriate quantity to quote in a measurement of mass loss was the energy absorbed by the sample rather than the irradiated (incident) energy. It is probably true that over a range of 0 - 100 nm thickness at 100 keV the absorbed energy varies about linearly with thickness as you suggest. However, it is only possible to compare data obtained at different beam energies and section thicknesses if the absorbed dose is used, since once the sample thickness exceeds one or two inelastic mean free paths at the incident energy the energy deposition is certainly not linear with mass thickness. Few biological "thin" sections, especially when examined in a probe at 50 keV or less, satisfy this criterion and consequently failure to correctly account for the fractional absorption of energy will lead to an apparent variation in the characteristic dose D 1/e for mass loss and an overestimate of its actual magnitude.

B.L. Gupta: As you know, doping plastic polymers with metals makes them electrically conductive. Does your observation about the lack of low temperature protection to beam damage and mass loss apply only to the sections of the standards or does it also apply to biological sections, particularly cryosections?

D.C. Joy: The observations I discussed apply only to these resins and in the context of our microscope (Hitachi H-800). Our experience has been that only when there is significant partial pressure of water vapor in the microscope do we see a change in the rate of mass loss between room temperature and liquid nitrogen conditions. When the partial pressure is low, because of bake-out of the column and the use of liquid nitrogen trapping, we see little difference for resin samples. This does not seem to correlate with the apparent conductivity of the resin.

Imaging and Microanalysis by Electron Spectroscopy

F.P. Ottensmeyer
Department of Medical Biophysics
University of Toronto and
the Ontario Cancer Institute
500 Sherbourne Street
Toronto, Ontario M4X 1K9
CANADA

The dominant strength of electron microsopy is the ability to produce high resolution images. For stained biological specimens in bright field this means a biologically interpretable detail of about 2 nm, limited by the Niquist sampling criterion referred to the size of the heavy metal ion stain. For thin unstained specimens, a compromise is struck between enough electron exposure for good counting statistics and a sufficiently low dose to limit radiation-induced structural alterations (Rose, 1977). Here an interpretable resolution of 0.3-0.5 nm has been reached (Ottensmeyer et al., 1975; Ottensmeyer et al., 1977; Andrews & Ottensmeyer, 1982). For individual heavy atoms and for crystalline specimens in materials science, the limit is the resolving power of the objective lens (Henkelman & Ottensmeyer, 1971; Crewe et al., 1970; Uyeda et al., 1978).

Spatial Resolution in Microanalysis

In biology, virtually all of our knowledge of structure and morphology at the subcellular level down to the fine structure of macromolecules has been derived from electron micrographs. The desire to add knowledge of chemistry to this structural information until recently has resulted of necessity in the acceptance of a loss in resolution in the localization of the analysis, and often, for single point measurements, in a loss in spatial dimensionality from 2D to zero. Cytochemical and autoradiographic techniques identify structures of interest to between 5,000 and 500 nm. Immunocytochemical approaches and electron probe x-ray microanalysis improve resolution to about 500 nm in general, although for overnight exposures the latter technique has produced elemental maps of phosphorus and calcium in a section of muscle tissue with a resolution of about 8 nm (Somlyo et al., 1981), still well above a factor of 10 worse than the resolving power of the microscope when imaging with electrons.

In contrast, the high resolution potential for imaging with electrons is also virtually completely realized for microanalysis using electron energy loss analysis (EEL analysis) both in the dedicated scanning transmission electron microscope (STEM) and in the fixed beam electron microscope (FBEM). The potential of 0.4 nm resolution or better in microanalysis has been calculated theoretically for low and high energy loss using quantum mechanical approaches (Kohl & Rose, 1985) and specifically for low loss plasmon excitations (Cheng, 1987). Experimental measurements had produced values between 0.3 to 0.5 nm over an energy range of 15 eV (plasmons) to 360 eV (inner shell excitations) (Scheinfein & Isaacson, 1986; Adamson-Sharpe & Ottensmeyer, 1981).

Sensitivity of Detection by Electron Energy Loss Analysis

The high spatial resolution capabilities of EEL imaging make the technique ideally suited to detect and identify a small number of atoms that occur naturally in highly concentrated and highly localized form. Thus, 30 to 50 atoms of phosphorus have been detected with a signal to noise ratio of 5 (Rose, 1977) in nucleoprotein particles and oriented phospholipid membranes using the FBEM (Harauz & Ottensmeyer, 1984; Adamson-Sharpe & Ottensmeyer, 1981). For uranium this limit has been extended to a cluster of 3 atoms (Mory & Colliex, 1988). Although these numbers of atoms are quite small, this sensitivity limit is achieved only because on the one hand, the local concentration is high (within the volume of the 2 nm double helix of DNA the phosphorus concentration is about 3 molar), and on the other, the spatial resolution of EEL analysis is sufficient to probe within that volume. For techniques with lower effective spatial (area or volume) resolution, such as occurs with thicker specimens, with EEL analysis using larger probe size, or with x-ray microanalysis, the specific signal from the same small number of atoms is diluted into the background signal from all the atoms in the larger volume.

The limit in the detection of the minimum mass fraction (minimum concentration) can nevertheless be quite good for a single point analysis with very careful treatment of the electron energy loss spectrum. Shuman et al. (1987) have detected a concentration of 2 mmol/kg calcium in a thin polyvinylpyrrolidine (PVP) film using signal averaging, matched PVP background subtraction, phase-locked differential spectral analysis, and matched filtering with a known calcium signal. This concentration corresponds to an atomic mass fraction of 25 parts per million, and the measured results extrapolate to a detection of 5 calcium atoms at a signal to noise ratio of 5 in a 10 nm cylinder (probe size) on a 25 nm thick PVP film using a 4.6 nA beam current for 100 sec. It is highly unlikely that such an analysis will be carried out point by point for elemental mapping.

Elemental Imaging

Imaging of the atomic elements with high spatial resolution and excellent sensitivity of detection is limited by the strength of the specific signal from the chosen element. In EEL analysis, this signal strength is high for a number of reasons. First, the energy loss event is the primary event in the electron-specimen interaction. In contrast, x-ray emission is one secondary consequence, the number of x-rays being limited beyond the primary event by the fluorescent yield, a number less than one, especially for low atomic number elements for which Auger emission predominates. Second, the scattering of the energy loss electrons is strongly forward-peaked, permitting a high collection efficiency (close to 1) for the analysing spectrometers. In contrast, x-ray detectors collect only about 1% of the isotropically emitted x-rays (Johnson et al., 1988). Third, electrons can be strongly focussed, not only to permit an additional matching of higher angle electron scattering with the spectrometer acceptance angle, but also to produce high resolution images or electron beam probes.

This combination of factors has meant that distributions of elements can be obtained with spatial resolutions equivalent to, or close to, the best electron micrographs. In our hands it has resulted in the delineation of the DNA-phosphorus supercoil within the 10 nm sub-unit of chromatin, the nucleosome (Bazett-Jones & Ottensmeyer, 1981, 1982; Harauz & Ottensmeyer, 1984). It has permitted the visualization of phosphorus in cross-sections of viral lipid membranes (Adamson-Sharpe & Ottensmeyer, 1981) and in the RNA of ribosomal particles (Korn et al., 1983). It has provided highly detailed maps of the relative concentrations and concentration ratios of calcium, phosphorus, and sulphur in mineralizing cartilage (Arsenault & Ottensmeyer, 1983) and of the distribution of Ti, N, O, S, Mn, and Cu in low alloy steels (Ottensmeyer et al., 1988).

Quantification in EEL Analysis

In spite of the strong demonstrated performance of the use of energy loss electrons in terms of spatial resolution and sensitivity of detection, absolute quantitative elemental analysis in EEL imaging and spectroscopy is still in its infancy. In principle, quantification is simple, the number of atoms "N" per unit area being given by:

$$N = \frac{\Delta I(\alpha,\varepsilon)}{I_T(\alpha,\varepsilon)} \times \frac{1}{\sigma(\alpha,\varepsilon)} \qquad (1)$$

where $I_T(\alpha,\varepsilon)$ is the total beam current coming into the spectrometer with acceptance angle "α", $\Delta I(\alpha,\varepsilon)$ is the net signal from the ionization of a characteristic inner shell of the specific atoms, obtained with an energy window "ε" on the spectral ionization edge, and $\sigma(\alpha,\varepsilon)$ is the partial ionization cross-section of the characteristic inner shell electron. In practice, this cannot yet be easily carried out.

The problem is two-fold. First, experimental absolute values for $\sigma(\alpha,\varepsilon)$ are very sparse. It is difficult to obtain reliable standards with known absolute concentrations of biologically important atoms. Second, the complexity of the fine structure of the ionization spectrum provides a wealth of potential information on the joint density of states and on solid state or chemical effects, but it makes it very difficult to calculate differential cross-sections from first principles with the accuracy needed for quantification. Theoretical cross-sections for K-shell ionization edges are the most reliable (Egerton, 1986), while discrepancies between measured and theoretical L cross-sections are still fairly large (Table 1) due to the simplifying

TABLE 1: Comparison of Experimental & Theoretical Scattering Cross-Sections

ELEMENT	EDGE	HYD* ($\times 10^{-20}$	H-S* ($\times 10^{-20}$ cm^2/atom	EXP* ($\times 10^{-20}$)	STANDARD
P	$L_{2,3}$	1.90	2.05	2.26 (150 eV)**	Poliovirus
P	$L_{2,3}$			2.68 (150 eV)**	$Ca_{10}(PO_4)_6(OH)_2$
Ca	$L_{2,3}$	0.31	0.43	0.49 (360 eV)**	$Ca_{10}(PO_4)_6(OH)_2$

* All results were calculated or measured under the same conditions: 80 keV incident energy,15 mrad half-angle, 15 eV energy window.

** Energy loss at which the cross-sections were measured.

HYD = Modified Hydrogenic Model (Egerton, 1986)
H-S = Hartree-Slater Model (P. Rez, personal communication)
EXP = Experimental Values

approximations in the theoretical approaches. Recently, solid state effects are being added to the approximations with very promising results (Weng & Rez, 1988) and attempts are being made to extend theory to higher order shells (Skiff et al., 1988). In the latter case, again, solid state effects are being neglected at this stage and results are matched to the shape of the experimental spectra only well above the edge.

The importance of greater accuracy in the calculation of theoretical cross-sections and the need for matching the solid state in the comparison of our measurements of the absolute phosphorus $L_{2,3}$ cross-section from poliovirus with theory (Table 1 and Fig. 1), as well as in the comparison of the difference in fine-structure (Fig. 2) near the same edge from the organic phosphodiester bonds of nucleic acids and the inorganic structure of hydroxyapatite ($Ca_{10}(PO_4)_6(OH)_2$).

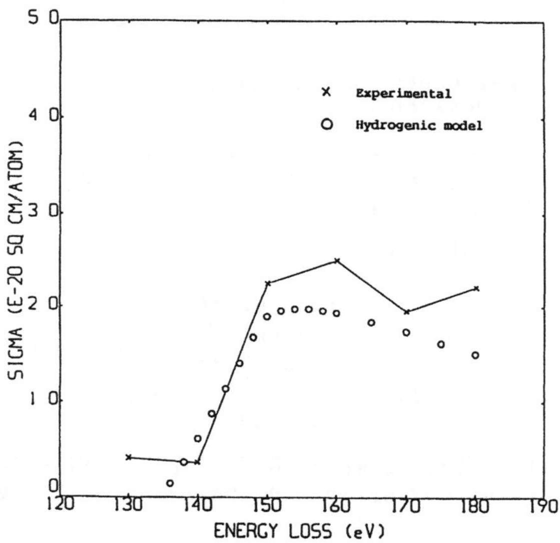

<u>Figure 1</u>. Phosphorus Cross-Sections. Experimental values from poliovirus, theoretical values from the modified hydrogenic model. Conditions as in Table 1

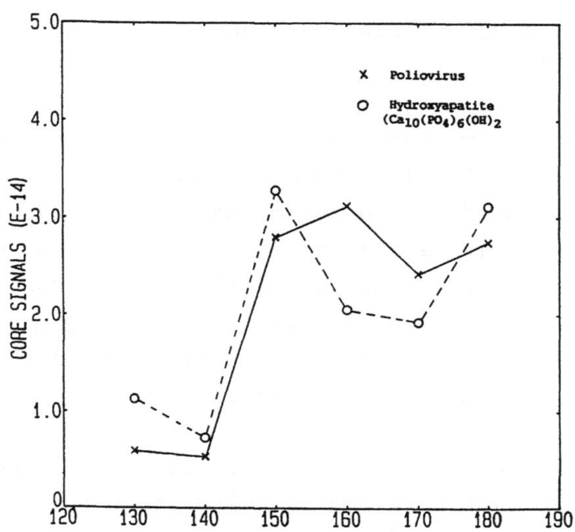

<u>Figure 2</u>. Phosphorus Cross-Sections. Experimental values from poliovirus and from hydroxyapatite. Conditions as in Table 1.

Relative experimental measurements of $\sigma(\alpha,\varepsilon)$ abound, primarily in the form of EEL spectra (e.g., Ahn & Krivanek, 1983). In addition, tabulations are becoming available of ratios of cross-sections measured with "ε" as large as 50 or 100 eV (Skiff et al., 1988; Leapman & Ornberg, 1988). Energy bands this large may be useful for simple compounds or alloys in materials science, or for well-separated K-edges, but for complex specimens such as generally studied in biology differential or partial cross-sections with about 15 eV resolution are required at least to prevent overlap of information from L-edges of nearest-neighbour atomic number elements. Nevertheless, the increasing scientific activity in this area over the last year indicates that these limitations are currently being addressed.

Other Limitations

The lack of absolute cross-sections is the most important present fundamental limit for quantification. In addition a number of limitations must be considered that prevent ubiquitous and indiscriminate application of the use of energy loss electrons for analysis. These concern specimen preparation:

1) Obviously, the specimen must not have lost or displaced the atomic element of interest before it is even placed in the electron microscope. The examination of quick-frozen hydrated specimens appears ideal, but only fairly thick sections of these specimens can be cut at present.

2) The specimen should not exceed a thickness of about 0.3 mean free paths for inelastic electron scattering in order to avoid interpretive difficulties due to multiple scattering. At 80 keV, this translates to a thickness of about 30 nm of carbon or about 40 nm of polymerized Spurr's embedding resin. Frozen hydrated cryosections of 100 nm or more are still too thick, although, if imaging but no micro-analysis is desired, then such sections can be examined with greater contrast using elastically scattered electrons only. Deconvolution techniques to address multiple scattering effects in thick specimens are being tried on energy filtered spectra. However, it is not likely that they will be easily applied to elemental mapping (Swyt & Leapman,

1982; Shuman & Somlyo, 1987; Bauer & Scholz, 1987).

3) Bulk specimens cannot be examined for EEL microanalysis, since the technique as practised at present requires forward scattered electrons that ideally have interacted with the specimen only once.

Auxilliary Advantages

Electron energy loss analysis enhances every mode of operation of the transmission electron microscope and adds a number of new possibilities. Elastic imaging, mentioned above, provides greater contrast in bright field, dark field and diffraction modes (Ottensmeyer, 1984). Energy selected imaging of thick sections with a low but non-specific energy loss provides modulation in contrast that is unique to the technique (Bauer, 1987). The availability of different combinations of signals, e.g., total, elastic, inelastic, permits imaging that can be made independent of thickness (Jeanguillaume & Tence, 1987) and in the limit of thin specimens can be proportional to Z, the atomic number (Crewe et al., 1970). Energy selected diffraction and angle dispersive spectroscopy are new approaches made possible by the technique (Reimer et al., 1988). Finally, since the spectrometers can be installed in currently unused locations in the electron microscope, there is not interference but rather complementarity with other auxilliary techniques such as x-ray microanalysis or cryomicroscopy.

Summary

Electron energy loss analysis in the electron microscope theoretically and experimentally can provide images of atomic distributions with spatial resolutions of 0.3 to 0.5 nm. Sensitivities of detection have been measured as good as 30 atoms of phosphorus and 3 atoms of uranium, while careful spectral analysis provided a limit of 2 mmol/kg or 25 ppm of calcium for the minimum mass fraction. Absolute quantification is still in its infancy, although relative quantification is well developed. Theoretical treatment of near-edge fine structure has begun to shed light on solid state or chemical effects.

For non-analytical applications, EEL filtering provides improved contrast via elastic imaging in bright field, dark field and diffraction. New modes of operation provide thickness independent imaging by signal combination, provide contrast modulation for thick sections by non-specific low-loss energy selection, and permit the study of angle and energy dispersion in novel combinations of diffraction and electron spectroscopy.

Acknowledgements

The work was supported by the Ontario Cancer Treatment and Research Fountation and by grants from the National Cancer Institute and from the Medical Research Council of Canada.

References

Adamson-Sharpe KM, Ottensmeyer FP (1981). Spatial resolution and detection sensitivity in microanalysis by electron energy loss selected imaging. J. Microsc. (Oxford) 122: 309-14

Ahn CC, Krivanek Ol (1983). EELS Atlas. Gatan Inc., Warrendale PA

Andrews DW, Ottensmeyer FP (1982). Electron microscopy of the poly-l-lysine alpha helix. Ultramicroscopy 9: 337-48

Arsenault AL, Ottensmeyer FP (1983). Quantitative spatial distribution of calcium, phosphorus and sulfur in calcifying epiphysis by high resolution spectroscopic imaging. Proc. Natl. Acad. Sci. USA 80: 1322-1326

Bauer R (1988). Electron spectroscopic imaging: An advanced technique for imaging and analysis in transmission electron mircroscopy. Meth. in Microbiol. 20:113-46

Bauer HD, Scholz W (1987). An approximate deconvolution for EELS quantification of thicker samples. Ultramicroscopy 23: 109-14

Bazett-Jones DP, Ottensmeyer FP (1981). Phosphorus distribution in the nucleosome. Science 211: 169-70

Bazett-Jones DP, Ottensmeyer FP (1982). DNA organization in nucleosomes. Can. J. Biochem. 60: 364-70

Cheng SC (1987). Localization distance of plasmons excited by high energy electrons. Ultramicroscopy 21: 312-13

Crewe AV, Wall J, Langmore J (1970). Visibility of single atoms. Science 168: 1338-40

Egerton RF (1986). Electron spectroscopy in the electron microscope. Plenum Press, New York

Harauz G, Ottensmeyer FP (1984). Nucleosome reconstruction via phosphorus mapping. Science 226: 936-40.

Henkelman RM, Ottensmeyer FP (1971). Visualization of single heavy atoms by dark field electron microscopy. Proc. Natl. Acad. Sci. USA 68: 3000-4

Jeanguillaume C, Tence M (1987). How to obtain a thickness-independent image in a STEM. Ultramicroscopy 23: 67-76

Johnson D, Izutsu K, Cantino M, Wong J (1988). High spatial resolution spectroscopy in the elemental analysis and imag-

ing of biological systems. Ultramicroscopy 24: 221-36
Kohl H, Rose H (1985). Theory of image formation by inelastically scattered electrons in the electron microscope. Adv. Electron. Electron Phys. 65: 173-96
Korn AP, Spitnik-Elson P, Elson D, Ottensmeyer FP (1983). Specific visualization of ribosomal RNA in the intact ribosome by electron spectroscopic imaging. Eur. J. Cell. Biol. 31: 334-40
Leapman RD, Ornberg RL (1988). Quantitative electron energy loss spectroscopy in biology. Ultramicroscopy 24: 251-68
Mory C, Colliex C. (1988). Elemental analysis at the single atom detection level by processing sequences of energy filtered images. In: Electron Beam Induced Spectroscopies with High Spatial Resolution. NSF/CNRS Workshop, Aussois, France, pp. 472-81
Ottensmeyer FP, Whiting RF, Schmidt EE, Clemens RS (1975). Electron microtephroscopy of proteins: a close look at the ashes of myokinase and protamine. J. Ultrastruc. Res. 52: 193-201
Ottensmeyer FP, Andrew JW, Bazett-Jones DP, Chan ASK, Hewitt J (1977). Signal to noise enhancement in dark field micrographs of vasopressin: filtering of arrays of images in reciprocal space. J. Microscopy 109: 259-68
Ottensmeyer FP (1984). Energy selecting electron microscopy. In: Electron optical systems for microscopy, microanalysis and microlithography. Hren J, Sewell P, Monroe E, Lenz F (eds). SEM Inc., AMF O'Hare IL, 245-51
Ottensmeyer FP,d Andres DW, Arsenault AL, Heng YM, Simon GT, Weatherly GC (1988). Elemental imaging by electron energy loss microscopy. Scanning (in press)
Reimer L, Fromm I, Rennekamp R (1988). Operation modes of electron spectroscopic imaging and electron energy loss spectroscopy in a transmission electron microscope. Ultramicroscopy 24: 339-54
Rose A (1977). Vision: Human and Electronic. Plenum Press, New York, p. 15
Scheinfein M, Isaacson M (1986). Electronic and chemical analysis of fluoride interface structures at subnanometer spatial resolution. J. Vac. Sci. Technol. B4: 326-32
Shuman H, Somlyo (1987). Electron energy loss analysis of near-trace-element concentrations of calcium. Ultramicroscopy 21: 23-32
Skiff WM, Carpenter RW, Lin SH, Higgs A (1988). K,L,M,N,O and P ionization cross-sections for electron energy loss spectroscopy. Ultramicroscopy 25: 47-60
Somlyo AV, Shuman H, Somlyo AP (1981). Microprobe analysis of biological systems. Academic Press, New York, p. 103-26.
Swyt CR, Leapman RD (1982). In: Scanning electron microscopy, Vol. 1, (ed.) Johari O, SEM Inc., AMF O'Hare IL, p. 737.
Uyeda N, Fujiyoashi Y, Kobayashi T (1978). Discrimination of individual atoms in molecular images of chlorinated copper phthalocyanine. In: Electron Microscopy 1978. 9th Intnl. Congr. Electr. Microsc., Toronto, pp. 242-3.
Weng X, Rez P (1988). Solid state effects on core electron cross-sections used in microanalysis. Ultramicroscopy 25: 345-48.

Conference Discussion

D.C. Joy: In your 3D reconstruction there were many missing angles. How does this affect the result? Can symmetry operations help? Are your projection angles unique?

P. Ottensmeyer: Orientations of the nucleosomes which are sparsely represented in the 3D reconstruction from 2D projections of these particles cause a larger point-spread function or equivalently a worse resolution in the direction of missing angles. Symmetry operations are always helpful in improving statistics if they can be justifiably used. We did not use such operations in order not to bias the outcome towards the eventually observed dyad symmetry.

Computer-assisted searches for the most likely single assignments for the nucleosomes projections resulted in over 50 unique orientation assignments. For only 4 nucleosomes was there an unresolved choice between two orientations. Both orientations were used in these cases in the reconstructions.

N. Roos: It is known from high resolution EM on protein crystals that doses of 1-10 e/A^2 can cause damage to the fine structure (range $<$10 A). You are exposing your specimen to $\leq$ 5000 e/A^2. How "real" are the models you are presenting and what is the spatial resolution you can obtain on for example nucleosomes?

P. Ottensmeyer: To answer the last part first, the spatial resolution in the nucleosome is about 10 A. However, in order to be able to state such a result in light of the crystal results, we have had to convince ourselves that the two-dimensional projection of the three-dimensional object is still imaged with biologically interpretable detail at electron exposures well above the 1-10 e/A^2. Since the image is formed from the two-dimensional mass distribution, the requirement on specimen damage is less stringent than for the three-dimensional crystal, permitting the collapse of structure in the third dimension within the 2D spatial resolution element before a loss in resolution results.

The spatial resolution in an energy loss image per se is not

limiting having been calculated theoretically and measured experimentally to be around 0.5 nm or lower (Kohl & Rose, Adv. Electron, Electron Phys. 65:173-196, 1985; Scheinfein and Isaacson, J. Vac. Sci Technol. B4:326-332 , 1986; Adamson-Sharpe and Ottensmeyer, J. Microsc. 122:309-314, 1981). Radiation-induced structural alterations then are the resolution-limiting factor in images of biological macromolecules.
For small peptide hormones the best resolution observed at exposures of 1000 e/A^2 was about 0.3 nm (Ottensmeyer et al., J. Microsc. 109:259, 1977; Ultramicroscopy 3:303, 1978) presumably since the small height of the peptides, 3-4 atoms at most, limited lateral diffusion. For protamine and adenylate kinase, larger proteins, 0.5 to 0.6 nm detail was observed (Ottensmeyer et al., J. Ultrastruct. Res. 52:193-201, 1975; Proc. Natl. Acad. Sci. 72:3953-3955, 1975) verified for protamine by X-ray crystallography (Warrant & Kim, Nature 271:130, 1978). Completely defined objective criteria and computer analysis were used to confirm the presence of the alpha-helical structure at 0.54 nm as well as remnants of the amino acid side chains in images of poly-L-lysine (Andrews and Ottensmeyer, Ultramicroscopy 9:337-348, 1982).
Thus, the potential spatial resolution of the detail in the nucleosomes is not out of line with resolutions found at similar electron exposures for other biological macromolecules. Moreover, the detail observed for histone H3 in our structure is remarkably similar to that described subsequently by Richmond et al. (Nature 311:532, 1984), while the overall shape agrees well with the protein core-particle in the crystal structure of Burlingame et al. (Science 228:546, 1985). A more extensive discussion of the reality of the nucleosome model was given by Erickson and Ottensmeyer (Science 233:1429-1431, 1986).

T.A. Hall: When one gets an X-ray elemental image in a section containing local microareas of very high mass thickness, one can get false positives in these spots if the background correction is imperfect. Can a similar effect occur and limit the reliability at low concentrations with your composite-image technique?

P. Ottensmeyer: The simple answer is yes. However, since electron energy loss analysis and imaging has a spatial resolution of close to 0.5 nm, it is less likely that there are local microareas within the analyzed unit area that have strong variations in mass thickness. This aspect is therefore of less concern for high resolution images but becomes important at lower spatial resolution. In addition, some form of background dependence with energy E, such as the analytic expression for the energy loss intensity $I = AE^{-R}$, is usually assumed. This expression is not correct in all cases, such as for the pre-edge dependence of Ca in combination with a carbon signal. The indiscriminate use of this expression can introduce positive or negative errors in elemental concentrations obtained by EEL analysis, whether it occurs in the STEM elemental image or in the FBEM. Current attempts to address the problem include correlation with standard spectra, well known in X-ray microanalysis, as well as convolution aproaches.

Application of X-Ray Microanalysis and Electron Energy Loss Spectroscopy to Studies of Secretory Cell Biology

Richard L. Ornberg[1] and Gemma A.J. Kuijpers
Laboratory of Cell Biology and Genetics
National Institute of Diabetes, Digestive and Kidney Diseases
National Institutes of Health
Bethesda, MD 20892

INTRODUCTION

The physical chemical mechanisms underlying secretion and exocytosis are unknown. Our current understanding relies on a plethora of biochemical studies of fractionated cells and isolated organelles (Poisner and Trifaro, 1982) and the validity of this approach has to be questioned until the in situ biochemical situation is known. For example, exocytosis is a sequence of structural events consisting of the mutual approach of the secretory granule and plasma membrane, membrane-membrane contact, exocytotic pore formation, and pore widening with the dissolution of stored contents into the external space (Ornberg and Reese, 1981). From studies of isolated granules from adrenal medullary chromaffin cells, it was proposed that the granules, through a ATP-driven proton and chloride influx, provide a chemiosmotic driving force for pore formation and exocytosis (Pollard etal, 1984). However, we simply do not have sufficient information regarding the microchemical composition of the "cellular test tube" to be certain that this does occur in the cell.

Research in our laboratory has used analytical electron microscopic methods on appropriately prepared rapid frozen tissue

---

[1]Baxter Healthcare Corporation, Baxter Technology Park, WG2-2S, Round Lake, IL 60073

to make in situ measurements of the chemical composition of cellular compartments. In this paper we will review recent developments in our study of the role of the secretory granules in the secretion of catecholamine from adrenal chromaffin cells.

TISSUE PREPARATION METHODS

In preparing tissue for analytical EM, one is immediately faced with a plethora of available methods ranging from chemical fixation and precipitation reactions (ie. calcium pyroantimonate) and analysis of plastic sections to quick freezing, cryoultramicrotomy, and analysis of frozen sections. Although many will cling to the simpler chemical methods, most serious investigators, being aware of the fixation artifacts (Crang and Klomparens, 1988) and accompanying ion redistributions, will choose quick freezing and cryoultramicrotomy. Freezing methods not only minimize artifacts but provide the capability of measuring a number of elements simultaneously in as close to the natural distribution as possible. The quick freezing methods for our work on suspension cultures of isolated bovine adrenal chromaffin cells have been described in detail (Ornberg, 1985, Ornberg etal, 1988). Cryosections of fresh frozen cells suspensions are cut on a Reichert Ultracut E microtome with FC-4 cryoattachment. Sections are sandwiched in formvar-carbon coated folding grids and are transferred into the electron microscope in the frozen hydrated state with a cryotransfer stage. They are either examined immediately for low dose EELS measurements or are freeze-dried in the column for xray analysis.

Ultrastructure of Cryosectioned Cells

Thin frozen hydrated sections cut from frozen cell suspensions invariably contained some small amount of compression. In frozen hydrated sections (Fig 1), compression appeared as a horizontal spaces running perpendicular to the cutting direction. Upon freeze drying, these spaces coincided with the cell boundary. The major factor influencing compression was

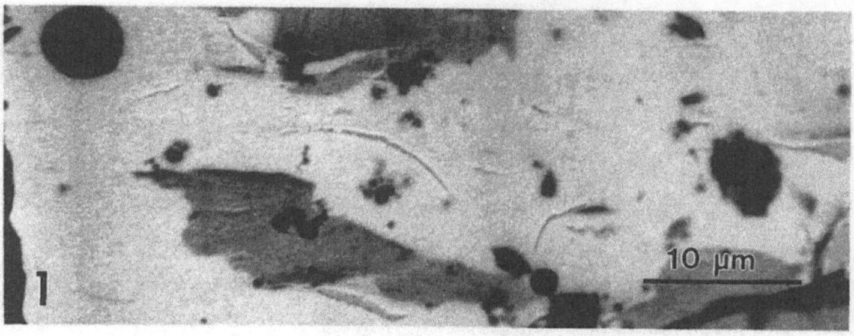

Figure 1. Dark field STEM micrograph of frozen hydrated section. Compression appears a small cracks at cell boundary (arrow). Overlying fragments of section and ice crystals appear dark (see text).

ambient temperature. Sections from the extreme surface layer cut best at the lowest temperatures, while from deeper and presumably less well frozen layers, a warmer temperature of approximately -120 °C was required. Antistatic charge neutralization also aided the reduction of compression although excessive discharging was particularly detrimental. Ultrastructural detail typical of rapid frozen, freeze substituted thin section views of chromaffin cells was observed in freeze dried cryosections imaged in dark field STEM (Fig 2). Nuclei, mitochondria, and the secretory chromaffin granules were easily identified. Mitochondria were distinguished by their cristae and their location in the perinuclear Golgi region. Chromaffin granules appeared as circular or oval shaped structures, ca 250 nm diam., with bright, electron dense coreswith an occasional dark, or electron translucent intragranular vesicles inside. Less well defined structures such as endoplasmic reticulum and Golgi apparatus were occasionally observed in the better sections.

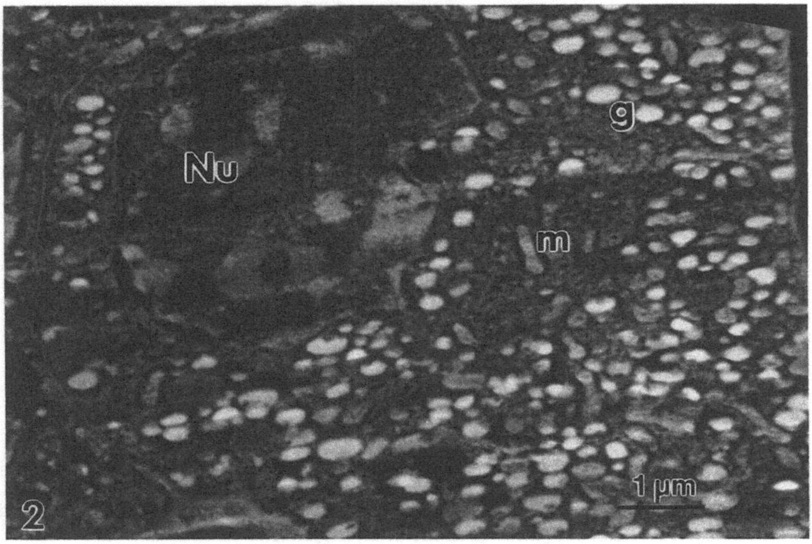

Figure 2. Dark field STEM micrograph of bovine adrenal chromaffin cell. Nucleus (Nu), mitochondria (m), and granules (g) readily observed.

EELS Determination of Mass Thickness and Water Content

The mass thickness of a thin cryosection is particularly important for microchemical analysis of biological cryosections in that if it is determined before and after dehydration, the water content of any given space can be estimated. The mass thickness of a thin sample can be determined from the amount of electron scatter. One method using the elastic scattered electron signal from an annular dark field detector requires an internal mass standard or a separate measure of incident beam current (Wall, 1979). However, this is not always possible with biological cryosections. Furthermore, with transmission electron microscopes equipped with scanning units and dark field detectors, the dark field signal can be very nonlinear with respect to mass thickness. Indeed, ice crystals lying on a thinner frozen section (Fig 1) appear as dark objects rather than light objects. Thus plural elastic scattering in the mass thick ice crystals results in scattering outside of the acceptance angle of the dark field detector, a loss on dark field signal

Alternatively, it is possible to measure mass thickness directly from the amount of inelastic scattering in the energy loss spectrum (Leapman and Ornberg, 1983, 1988). For this, the unscattered (zero loss) intensity, I_o, and the total incident electron flux (total counts in spectrum), I_{tot}, can be used in the following expression for mass thickness;

$$\rho t = (M/N_o\sigma_{in})\ln(I_{tot}/I_o) \qquad \text{Eq. 1}$$

where M is the mean molecular weight, N_o is Avogadro's number, and σ_{in} is the average inelastic cross section. Advantages of the EELS method are that it does not require internal standards; it can be performed with a very low incident dose ($\leq 10\ e^-/Å^2$) and it is linear for samples up to 300 nm thick (Leapman etal, 1987). From measurements of mass thickness before and after dehydration and with the mass fraction of the support film subtracted, the fraction of water is given by;

$$f_{H2O} = 1 - \rho t_{dry} / \rho t_{hyd} \qquad \text{Eq. 2}$$

where "hyd" and "dry" refer to hydrated and dehydrated. It is assumed that the hydrated section and support film have a uniform thickness. In testing this method on frozen sections of adrenal chromaffin cells, the chromaffin granule had a water mass fraction of 66% ± 5%. This value is in excellent agreement with biochemical measurements of a 60% to 70% water content for isolated granules. However, in further work on standard mixtures of water and various concentrations polyvinylpyrrolidone, PVP, two corrections had be made, Table 1 (Leapman and Ornberg, 1987). First, the redistribution of mass (or shrinkage) during dehydration had to be accounted for. For this, we simply measured a linear shrinkage factor, s, from sequential images of the section during the dehydration, and corrected the mass thickness for areal shrinkage (col 3 (A & B), Table 1). This correction provided a reasonable but systematically higher than nominal estimate of the dry mass fraction, $\rho t_{dehyd}/\ \rho t_{hyd}$ (col 5 (A), Table 1). Second, it became apparent that the average inelastic scattering cross-sections of the hydrated and dehydrated sections were not the same. Using the known stoichiometry of PVP and water and the atomic Hartree-Fock inelastic cross-sections previously reported (Leapman etal, 1983), the quantity, $M/N_o\sigma_{in}$, was calculated for water + PVP, 1.07 x 10^{-5} gm/cm^2 and for PVP, 0.8 x 10^{-5}gm/cm^2. After, multiplying the original relative mass data by the ratio of the PVP value to the water + PVP term, a near perfect agreement between the nominal and measured dry mass was obtained (col 6 (B), Table 1). In doing a similar calculation for the adrenal chromaffin granule, a water content of 58% was obtained.

Table 1. DRY MASS FRACTIONS (DMF) OF PVP SOLUTIONS AS DETERMINED BY EELS

Nominal DMF	Shrinkage	$\rho t_{dry}/\ \rho t_{hyd}$		Actual DMF	
		(A)	(B)	(A)	(B)
0.02	.314	0.033	0.025	0.015	0.011
0.10	.306	0.259	0.194	0.125	0.093
0.20	.211	0.436	0.326	0.271	0.202
0.40	.129	0.698	0.522	0.530	0.396

(A) $\sigma_{in}(dry) = \sigma_{in}(hyd)$ (B) $\sigma_{in}(dry) \neq \sigma_{in}(hyd)$

Light Element Analysis by EELS

Meaningful measurements of the light element distributions in cryosectioned cells by EELS may seem futile because the existing mixtures of protein, carbohydrate, and lipid have such similar carbon, nitrogen, and oxygen composition. However, if sufficient biochemical data describing the chemical composition of an organelle is known, knowledge of the light element composition can be useful to determine organelle variability _in situ_. Such is the case with the adrenal chromaffin granule. Based on numerous determinations (see Winkler etal, 1986), the chromaffin granule contains 120 mM -150 mM adenosine triphosphate (ATP), 550 mM catecholamine, 15 - 30 mM Ca and 220 mg/ml protein. From the known stoichiometry of ATP and catecholamine and the amino acid composition of chromogranin A, a major protein in the granule core, an expected percentage distribution of light elements, C,N,O,P,S, was formulated for the granule contents. This expected distribution indicated that 38 % of the total nitrogen of the granule core is associated with nucleotide, 33 % with protein and 39 % with catecholamine. By assuming that all of the phosphorous in the core is associated with nucleotide, every three phosphorous atoms in the light element distribution should account for five granule nitrogen atoms or 38% of the total granule nitrogen. Using ATP as a standard, the ratio of the relative cross section of the phosphorus K edge and nitrogen K edge was determined for the EELS spectrometer parameters. From this ratio and the phosphorus edge count from granule spectra, the nitrogen count and percent total phosphorus associated nitrogen was measured to be 33 $\pm$ 8% for 80% of the granules measured. The surprising result was that the remaining 20% of granules had a lower phosphorus to nitrogen ratio. One possible explanation for this is that these granules did not contain catecholamine but were immature "purinergic" granules. We are unaware of any biochemical evidence supporting this conclusion however and further analysis is required. Regardless, this result illustrates the potential power of EELS to biochemically analyze single granules <u>in situ</u> and suggests that light element analysis by EELS can play a role in elucidating secretory granule function and biogenesis.

X-ray Analysis of Chromaffin Cells and Granules

X-ray analysis of cell in frozen sections extends the microchemical analysis of single organelles to the ionic elements (Z $\geq$ Na) with considerable accuracy provided valid calibration and corrections are employed. Freeze-dried frozen sections of chromaffin cells were analyzed as previously described (Ornberg etal, 1988). The results our initial study of the <u>in situ</u> contents of chromaffin cell organelles and in particular the chromaffin granule have been described. The major finding of this work is that during routine isolation, granules undergo a profound redistribution in ion or element content. Indeed the final of steady state ion content appeared to depend on the Na and K content of the isolation buffer.

To compare further our analysis with previous wet biochemical measurements and to use it for a more physiological understanding of the granule, we determined water content and the wet weight elemental concentration in each compartment from the background xray counts as previously described. The granule, which served as an internal standard, had a water content of 66% (as determined by EELS), mitochondria had 70%, the nucleus had 82%, the cytoplasm had 88%, and the external medium had 89%. Assuming that all of the compartmental water was available for dissolving the elements, the element concentrations in units of mM/kg water (not tissue wet weight) was determined and is expressed in Table 2.

Table 2. Elemental Composition of Chromaffin Cell Compartments

Compartment	Concentration (mM/Kg water)						
	Na	Mg	P	S	Cl	K	Ca
Saline	152	2	2	32	128	6	3
Mitochondria	1	21	116	35	33	165	0.5
Nucleus	3	10	104	20	37	148	0.2
Cytoplasm	5	10	110	20	26	120	1.2
Granules	0	5	347	55	49	83	8.8
Isolated Granules:							
Na-MES Buffer	29	2	8	35	5	2	0.6
Granules	29	0	416	55	22	19	7.0
K-MES Buffer	2	0	12	58	7	26	1.0
Granules	8	0	418	61	27	40	12

The validity of this approach is supported by the close correlation of our wet weight concentration with previous reports of granule composition and with the known physiology of the cell. With the exception of Cl, which maybe lost during analysis, and

P , which is secreted by the cells, the ionic concentrations of the external medium agree with the saline formulation (150 mM NaCl, 5.6 mM KCl, 2.2 mM $MgCl_2$, 1.8 mM $CaCl_2$, and 15 mM HEPES[2] - NaOH buffer, pH 7.3). Furthermore, the K^+ diffusion potential of -75 mV which we calculated from the Nernst equation using 6 mM external K and 120 mM internal K is in reasonable agreement with measured membrane potentials of -65mV (Ornberg etal, 1988). Finally since the granule membrane is impermeable to phosphate and that all of the granule phosphorus is in the form of ATP we would predict from table 3 that the ATP content of our granule is approximately 115 mM which is in excellent agreement with previous biochemical measurements of 115 to 150 mM (see Winkler and Carmichel, 1982).

From our estimate of chloride, an ion more likely to be unbound and completely dissolved, a broader understanding of the physiological state of the granule emerges. Chloride permeates the granule membrane and has been shown to partition across the granule membrane according to the granule membrane potential (Pollard etal, 1976). Using the ratio of the cytosol and granule chloride concentration, 26 mM and 49 mM respectively, in the Nernst equation, we obtain a granule membrane potential of 16 mV inside positive. This value is much less than the 78 mV value obtained by Baker and Knight (1982) from thiocyanate distributions in permeablized cells. However this latter value was measured in the absence of permeant anions and should be considered a high estimate. For isolated granules, it is known that in the absence of an energy source such as ATP, the granule membrane potential is set by the external pH and the internal granule pH which is strongly buffered at pH 5.5 - 5.7 (Pollard etal., 1979). With the pH of our isolation buffer at 6.0 and thus a pH gradient of 0.5 units, we would expect a membrane potential of ca. 30 mV inside positive. Again this is in reasonable agreement with 34 mV and 37 mV calculated from chloride gradients in MES[3]-NaOH and MES-KOH buffer, respectively.

---

[2]HEPES, N-2-hydroxyethylpiperizine-N'-2-ethanesulfonic acid

[3]MES, 2-(N-morpholino)ethanesulfonic acid

Further examination of granules in defined medium and treatment conditions will allow us to assess the accuracy of measurements of membrane ion gradients for estimating membrane potentials.

Discussion; Current Status and Future Developments

Our work on chromaffin cells and granules, like the other fine work presented at this conference, has attempted to assemble a group of reliable preparative techniques and electron probe methods to make accurate and physiologically relevant measurements of elements in cells.

Of the techniques involved in tissue preparation, ie. tissue freezing, cryoultramicrotomy, and section freeze-drying, tissue freezing is clearly the most important. The freezing method should render the tissue free of detectable ice crystals at the resolution of the electron microscope. Such freezing is readily attainable with liquid helium cooled metal block freezing devices and, depending on the tissue, becomes exceedingly more difficult with liquid nitrogen cooled metal or liquid coolants. The minimization of ice crystals is important for two basic reasons. First, ice crystals contain only water and thus grow by the exclusion of the biological molecules and ions which we hope to measure in as close to the natural distribution as possible. Second, the growth of ice results in water and mass redistribution which preclude meaningful determination of water and wet weight ion contents. To our knowledge, no recent study has examined the issue of ice crystal size versus elemental content. Clearly, the fine structural ion distributions within secretory granules, endoplasmic reticulum, mitochondria on down to the level of the phospholipid membrane will require distortion free specimens.

Cryoultramicrotomy has become much less burdensome with recent advances in instrumentation. The microtomy of thin, flat, uncompressed sections from fresh, uncryoprotected frozen tissue is also simplified by ice crystal free freezing (McDowall etal, 1982). Quick frozen suspensions of cultured cells in saline have been the most difficult to cut. Although very good sections are usually attainable, a small, but detectable amount of compression is almost always observed. We have not found conditions which consistently eliminate this but are hopeful that recently developed diamond knives that have narrow included angles (Jesior,1989) will be helpful.

Section drying prior to xray analysis has been simplified by temperature controllable cryotransfer stages. The need for drying inside the vacuum of the EM column with such a stage has been demonstrated for certain tissues (Hagler, 1984) but is still not widely used. The major advantage that sections dry in an environment which minimizes section rehydration and element redistribution is opposed by the fact that a large amount of scope time is lost to sample preparation. Perhaps more important than this however should be methods to minimize electron probe induced mass loss. Lowering specimen temperature to -100 °C or lower has been shown reduce mass loss (Cantino etal, 1985). Lowering the specimen temperature can easily result in specimen rehydration and etching if the microscope vacuum is not properly maintained. One alternative and complementary solution, maybe to coat the hydrated section with carbon prior to or after drying similar to coating freeze-etch specimens. This would provide a conductive barrier to molecular species that have been fragmented by the electron beam and would otherwise escape the specimen. Thus the area of section preparation immediately prior to analysis may benefit from future developments.

In conclusion, we have attempted to obtain, at the highest resolution possible, information about the chemical composition of subcellular organelles in chroamffin cells. We have extended the results of conventional microanalysis to chemical concentrations to undertand the physiological function of the organelles <u>in situ</u>. We have also attempted a meaningful explanation of light element compositions obtained by EELS and can conclude that if enough biochemical information is available, meaningful correlations can be made and subclassifications of chroamffin granules can be determined. Although our results and methods thusfar are interesting, they are just a beginning of what we hope will lead to new discoveries in secretory cell physiology.

REFERENCES

Baker PF, Knight DE (1982) The Chromaffin Granule Proton Pump and Calcium-Dependent Exocytosis in Bovine Adrenal Medullary Cells. J Membr Biol 68:107-140

Cantino MF, Goddard LE, Wilkinson LE, Johnson DE (1985) In; Bailey GW (ed) Proc. 43st Annual EMSA Meeting. San Francisco Press, San Fransico, CA, 470

Crang RFE, Klomparens KL (1988) Artifacts in Biological Electron Microscopy Plenum Press, New York, N.Y.

Hagler HK (1984) Cryotransfer of Ultrathin Cryosections for Xray Microanalysis. In: Romig AD, Goldstein JI (eds) Microbeam Analysis - 1984. San Fransico Press, San Fransico, CA. 290-294

Jesior JC (1988) Use of Low Angle Diamond Knives Leads to Improved Ultrastructural Preservation of Ultrathin Sections. In: Albrecht RW, Ornberg RL (eds) The Science of Biological Specimen Preparation for Microscopy and Microanalysis 1988. Scanning Electron Microscopy, Chicago AMF O'Hare, IL, in press

Leapman RD, Ornberg RL (1987) Measurement of Water Content in Hydrated Sections by EELS In: Bailey GW (ed) Proc. 41st Annual EMSA Meeting. San Francisco Press, San Fransico, CA, p.590

Leapman RD, Fiori CE, Swyt C (1983) Mass Thickness Determination by Electron Energy Loss for Quantitative X-ray Microanalysis in Biology. J Microscopy 133:239-253

Leapman RD, Ornberg RL (1988) Quantitative Electron Energy Loss Spectroscopy in Biology. Ultramicroscopy 24:251-268

McDowall AW, Chang JJ, Freeman R, Lepault J, Walter CA, Dubochet J (1983) Electron Microscopy of Frozen Hydrated Sections of Vitreous Ice and Vitrified Biological Samples. J Microscopy 131:1-9

Ornberg RL (1985) Cryoultramicrotomy and Electron Microanalysis of Isolated Bovine Adrenal Chromaffin Cells. In: Muller M, Becker RP, Boyde A, Wolosewick JJ (eds) The Science of Biological Specimen Preparation for Microscopy and Microanalysis 1985. Scanning Electron Microscopy, Chicago AMF O'Hare, IL. 135-137

Ornberg RL, Reese TS (1981) Beginning of Exocytosis Captured by Rapid-freezing of Limulus Amebocytes. J Cell Biol 90:40-54

Ornberg RL, Kuijpers GAJ, Leapman RD (1988) Electron Probe Microanalysis of the Subcellular Compartments of Bovine Adrenal Chromaffin Cells. J Biol Chem 263:1488-1493

Pollard HB, Shindo H, Creutz CE, Pazoles CJ, Scott JH, Zinder O, Hotchkiss A (1984) An Osmotic Mechanism for Exocytosis from Dissociated Chromaffin Cells. J Biol Chem 259:1114-1121

Pollard HB, Shindo H, Creutz CE, Pazoles CJ, Cohen JS (1979) Internal pH and State of ATP in Adrenergic Chromaffin Granules Determined by ^{31}P Nuclear Magnetic Resonance. J Biol Chem 254:1170-1177

Pollard HB, Zinder O, Hoffman PG, Nikodejevic O (1976) Regulation of Transmembrane Potential of Isolated Chromaffin Granules by ATP, ATP Analogs, and External pH. J Biol Chem 251:4544-4550

Poisner AM, Trifaro JM (eds) The Secretory Granule. Elsevier Biomedical Press, Amsterdam

Wall J (1979) In: Johari O (ed) Scanning Electron Microscopy 1979. SEM Inc, AMF O'Hare, IL. 291

Winkler H, Carmichel SW (1982) The Chromaffin Granule. In: Poisner AM, Trifaro JM (eds) The Secretory Granule. Elsevier Biomedical Press, Amsterdam. 1-80

Winkler H, Apps DK, Fischer-Colbrie R (1986) The Molecular Function of Adrenal Chromaffin Granules: Established Facts and Unresolved Topics. Neuroscience 18:261-278

Conference Discussion

R. Reichelt: You mentioned in your talk that the characteristic EELS signal behaves linear up to higher thickness compared to the elastic darkfield signal acquired by an annular detector (AD). I do not really understand that. Assuming a collection angular range of the AD of say 15 to 150 mrad multiple scattering hardly gives scattering angles >150 mrad, i.e. the AD signal remains unchanged. In contrast to this characteristic inelastic events are removed from the selected energy window and/or the angular collection range of the spectrometer with a high probability, if an additional inelastic or elastic scattering event takes place, i.e. the characteristic event is not registered. Can you comment on that?

R. Ornberg: What you say is correct for the "dedicated" STEM. My perhaps overstated comment is limited to my experience with the TEM equipped with a scanning unit. These instruments are electron optically more complex and perhaps less well defined than the STEM you refer to. In the Hitachi TEM-STEM used here, you can see the nonlinearity in the contrast of the darkfield-hydrated image. Areas which contain fragments of section debris and are thus approximately two sections thick (ca 4 inelastic mean free paths) are not incrementally bright but have a reversed contrast. This is the nonlinearity that I was referring to. As for the EELS inelastic measurement, it has been shown to be linear out to four mean free paths and beyond (Leapman et al., 1984). I can only speculate that the nonlinearity we observe is due to the limited acceptance angle of the darkfield detector and the optical design of the TEM-STEM.

P. Ingram: Have you attempted to measure water-content maps using zero loss or low loss EELS?

R. Ornberg: No, we have not. However, Richard Leapman and colleagues have been working on an EELS parallel detection system for this purpose (see Leapman, this volume).

J. Shelburne: Have you had the opportunity to do electron probe microanalysis on the dense core granules of carcinoid tumors?

R. Ornberg: No. We have some data on pancreatic B-cells, fresh frozen liver, and mast cells, but no tumor cells.

N. Roos: You assume to have 60 nm thick cryosections of uniform thickness. What are the criteria for uniform thickness? How do you measure it? Compression cannot be avoided sectioning on a dry knife, how will compression affect the uniformity of the thickness?
R. Ornberg: Our criteria for checking hydrated section uniformity is simply the intensity or evenness of the contrast of the darkfield image. A more accurate measure would involve zero loss and whole spectrum images from which the inelastic mean free path could be calculated for each pixel. We have not been able to do that yet. The fine compression appears to be uniform, based on the ultrastructure we have observed. If and when organelles are compressed more than others, this artifact will negate the uniform thickness assumption and more corrections will have to be employed.

Th. v. Zglinicki: 1. Your sulfur values seem to be rather low. Did you measure at room temperature? 2. You found deviations of up to 20 % using your water content measurement technique even on "corrected" standards. I do not see the advantage of this method over classical EM methods using brightfield contrast.
R. Ornberg: 1. The temperature of our specimen varied from -60 to -75°C and we did notice some loss of S and Cl. 2. The advantage of the EELS measurement of water content is that it requires no internal standard, no beam current calibration, and photographic film measurement. The two spectral quantities needed require no sophisticated fitting routines or spectral processing.

R.A. Steinbrecht: Considering that there is still quite some variation in your data, have you ever compared the variation of element ratios with that of absolute concentrations? If the former were much smaller than the latter, this would indicate that background correction could still be improved.
R. Ornberg: Yes. We have calculated Ca/P and Ca/S ratios for

granules in situ and find a variance or standard of the mean which is not different from the absolute measurement. We have some preliminary evidence that there are at least two populations of granules in cells based on their Ca content and, therefore, believe that there may be a biological reason for the large standard error of the mean in some of our measurements.

T.A. Hall: You have discussed choosing among several alternative methods for measuring "mass thickness" (i.e. mass/area). Linders* has proposed and tested the use of the intensity of backscattered electrons. This measure should be linear up to very high mass/area and insensitive to inhomogeneity on a microscale. Do you think it is suitable?

*Linders PWJ and Hagemann P (1983), Ultramicroscopy 11:13-20.

R. Ornberg: The backscattered electron, BSE, intensity method is one which we have not tested but is analogous to the EELS method in that the measured signal is derived from a scattering event. However, I believe the backscatter signal will be much smaller than the EELS signal due to the low cross-section of the elastic backscatter event and the small acceptance angle of the backscatter detectors. Thus, for specimens, i.e. freeze-dried frozen or plastic sections, which can withstand the high dose needed to raise the signal to noise, the BSE method seems very suitable for measuring mass thickness. I would expect that rigorous use of this method would require some correction for variations in atomic composition of some cell compartments.

L. Edelmann: According to own experience it is necessary to freeze-dry cryosections (of muscle) rather a long time at low temperatures in order to reduce the linear shrinkage during freeze-drying to about 6 % (Edelmann, Scanning Electron Microsc. 1986 IV:1337-1356). Did you try long freeze-drying times at low temperatures? I would expect that you eventually may get a rather uniform small shrinkage of your cryosections.

R. Ornberg: No, we have not examined that. We were primarily concerned with removing as much water as possible to avoid subsequent mass loss during analysis.

X-Ray Microanalysis of Freshly Isolated Cells in Suspension

Alice Warley
Divisions of Anatomy and Biochemistry
U.M.D.S. St. Thomas's Hospital Medical School
Lambeth Palace Road, London SE1 7EH

Introduction

There is increased interest in the use of cells which have been isolated from tissues as experimental systems. The advantages of such preparations over the intact tissue have been listed as: ease of manipulation of the experimental environment; unambiguous interpretation of results; ability to study individual cell types; and the opportunity to observe time dependent processes by sampling from the preparation (Dow et al., 1981a; Buja et al., 1985; Lechene, 1986). The study of elemental concentrations in isolated cells should help in gaining an understanding about the role of different elements in the control of cell functions, and about alterations that occur in disease processes.

Cryofixation and X-ray microanalysis have been used to study diffusible elements in cells in culture and in freshly isolated cells (reviewed in Wroblewski and Roomans, 1984; Warley, 1987; Zierold and Schafer, 1988). The majority of these investigations have been on adherent cells growing as monolayers which have been studied whole (e.g. James-Kracke et al., 1980; Zierold, 1981; Ceder et al., 1982; Lechene, 1986). However if whole cells are used it is not possible to study subcellular organelles without interference from the surrounding cytoplasm. In addition the results from analysis are complicated by the presence of medium overlying the cells. The problems encountered in the preparation of whole cells for X-ray microanalysis have been well documented (Zierold 1981; Wroblewski et al., 1983; Lechene, 1986; Zierold and Schafer, 1988). These problems can be overcome if the cells are sectioned before analysis. Cryosectioning of adherent cells has been undertaken by Zierold et al. (1984) who sectioned cells growing on freeze etch supports. Cells have also been grown on collagen (Buja et al., 1985), Formvar (Tvedt et al., 1988)., or a gas permeable membrane (Zierold and Schafer, 1988) all of which can be sectioned.

Cells which are grown, or are isolated, in suspension can be studied by X-ray microanalysis after a simple preparation procedure consisting of centrifugation to form a thick cell pellet which is quench frozen to give a specimen suitable for cryosectioning. In this way sections of the cells can be studied in the presence of the medium in which they are prepared or grown, and analysis of cellular organelles can be undertaken. The technique of X-ray microanalysis of cells in suspension has been used successfully to study cells of the blood (Jones et al., 1979; Kirk et al., 1983; Lew et al., 1985). Although the elemental composition of intact heart tissue has been well studied using techniques such as flame photometry, atomic absorption spectroscopy (Polimeni, 1974; Poole-Wilson and Cameron, 1975; Ward and Cameron, 1984) and X-ray microanalysis (von Zglinicki et al. 1986), there have been few studies on isolated cardiac myocytes. In the present paper results from studies of cardiac myocytes will be presented and discussed as an application of X-ray microanalysis to the study of freshly isolated cells in suspension.

Methods for X-ray Microanalysis

Freshly isolated myocytes were concentrated by centrifugation at 500 g for 5 min, the supernatant was removed and drops of the thick cell suspension suspended on silver stubs and frozen by plunging into liquified propane using a Reichert KF80 freezing system. Cryosections were cut at 140 K using a Reichert FC4D cryoultramicrotome. The sections were transferred onto Ni grids and the grids transferred to a freeze drier in a brass block. The construction of the block was modified from Wendt-Gallitelli and Wolburg (1984). Sections were freeze dried overnight and carbon coated before analysis.

Analysis was undertaken on a Jeol 100C electron microscope in STEM mode using a beam current of 1.5 nA measured with a Faraday cage. Quantitative data was derived using the Quantem program supplied by Link Analytical (for further details of the analysis system see Spencer et al., 1988).

Results and Discussion

Methods for the isolation of cardiac myocytes are now well established (Dow et al., 1981 a,b). The suspension of myocytes obtained after isolation by perfusion of the heart with collagenase consists of two cell types with different morphology (Fig 1). It is

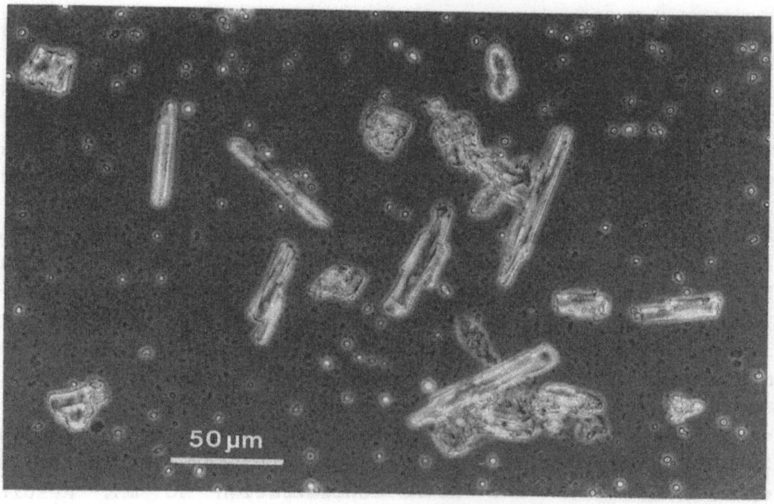

Fig. 1. Light micrograph of myocyte suspension showing round and rod shaped cells.

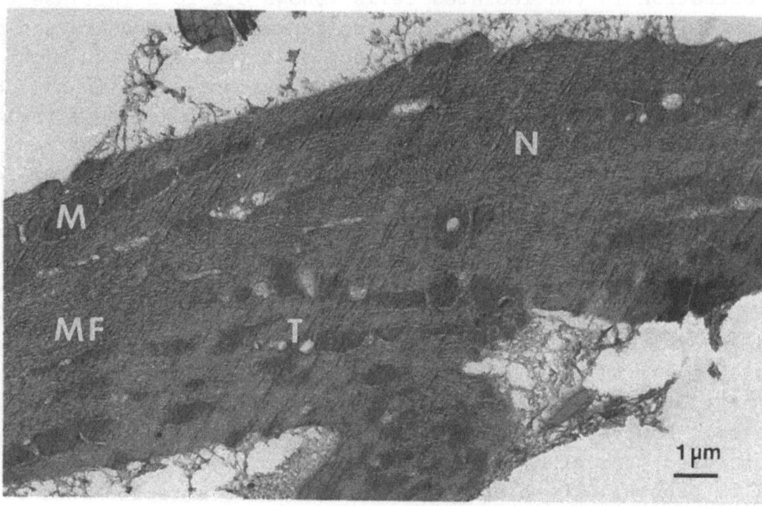

Fig. 2. Freeze dried cryosection of an isolated cardiac myocyte. N Nucleus; MF Myofibrils; M mitochondria; T T-Tubule.

generally agreed that the rod shaped cells are viable and show characteristics of intact cardiac tissue, whereas the round cells are non viable (Dow et al., 1981a). The results from X-ray microanalysis of the isolated cells show that elemental concentrations differ in the round and rod shaped cells. The round cells have high concentrations of Na, low concentrations of K, and very high concentrations of Ca (250 mmol/kg dry weight) in their mitochondria.

A freeze dried frozen section of a rod shaped myocyte is shown in Fig 2. Mean concentrations derived from X-ray microanalysis of isolated rabbit cardiac myocytes (rod shaped cells only) were; Na, 128 mmole/kg dry weight; Mg, 36 mmole/kg dry weight; and K, 407 mmole/kg dry weight. The X-ray microanalysis results were obtained from analysis of myofibrillar region of the cells. If the dry weight concentrations obtained from X-ray microanalysis are converted to mM concentrations using a water content of 76% for the myofibrils (von Zglinicki and Bimmler, 1987) the concentration of K in the isolated myocytes is 128 mM and the Na concentration 40 mM. Results for intracellular concentrations of elements obtained from bulk analysis of rabbit ventricle (Ward and Cameron, 1984; Ward, 1988) were Na, 20-30 mM; K, 125 mM; and Mg, 40 mmol/kg dry weight. The results for K and Mg from the two techniques are in good agreement. The mean Na concentration in the isolated cells shows a 2-fold increase compared to the tissue values. The increase in mean Na concentration in the isolated cells is due to the presence of some cells with high values rather than an overall increase in Na concentration in all of the cells (see Fig. 3). Concentrations of K also showed some variability in the individual cells. Thus X-ray microanalysis of the rod shaped cells shows a population which is heterogeneous in elemental concentrations.

When cells are isolated from intact tissues it is inevitable that damage will occur to some cells during the isolation procedure. Isolated cells which are capable of division are generally grown as monolayers and allowed to establish themselves in culture for varying periods of time before analysis (e.g. Ceder et al., 1982; Wroblewski et al., 1983; Abraham et al., 1985; Buja et al., 1985). Thus, in such preparations any changes caused by the isolation procedure are negated. However, when cells are isolated in suspension they are often used for study immediately and a number of non-viable cells will be present. Therefore there is a need to establish the viability of individual cells. Severely damaged isolated cardiac myocytes are readily apparent because of their round form. Elemental concentrations in the round cardiac myocytes are in agreement with the many X-ray microanalytical studies (Gupta and Hall., 1978; Somlyo et al., 1979; James-Kracke et al., 1980; Osornio-Vargas et al., 1981; Wroblewski and Edstrom, 1984; Buja et al., 1985; LeFurgey

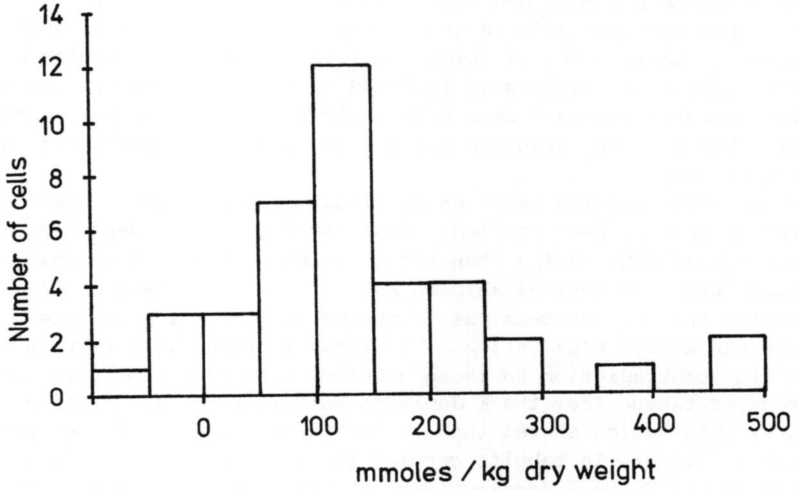

Fig. 3. Histogram showing distribution of values for Na concentration in individual myocytes.

et al., 1986) which show that non-viable cells are characterised by low K/Na ratios and by high levels of Ca in the mitochondria.

In addition to the non-viable cells a number of the rod shaped cells show increases in Na concentration. The higher concentrations could be due to damage to the cell membrane and/or alteration in cell volume. Higher concentrations of Na have been described in freshly isolated rat thymocytes (Warley et al., 1986; Warley, 1986) even after a relatively mild isolation procedure. The concentration of Na fell after the cells had been incubated for 1 h (Warley, 1986). The need for incubation of cells in a suitable medium to allow stabilisation after isolation has also been documented for isolated renal tubules (LeFurgey et al., 1986) and for cells isolated from rat renal papillae and rat liver (Zierold and Schafer, 1988). Further studies will show whether Na concentrations in isolated myocytes decrease after incubation.

Several studies have now been carried out on isolated cells (Hook et al., 1986; LeFurgey et al., 1986; Warley, 1986) in which results from X-ray microanalysis are compared with results from bulk analyses or ion-sensitive microelectrode analysis (Acker et al., 1985) of similar preparations. In all of these studies the results obtained by the different techniques were comparable showing that the technique of X-ray microanalysis is complimentary to other established techniques. Apart from the higher values for Na, the results obtained from X-ray microanalysis of isolated cardiac myocytes are consistent with results obtained from bulk analysis of this tissue, suggesting that the isolated myocytes are a good model for myocytes in the intact tissue.

Isolated cardiac myocytes obtained from rabbits on a K depleted diet have also been studied. Myocytes from the K depleted animals have higher K/Na ratios than those isolated from control animals. In cells from the control animals 25% of the cells have a K/Na ratio greater than 5, whereas 42% of the cells from the K depleted rabbits have the higher K/Na ratios. The X-ray microanalysis results seem to be in contradiction to those obtained from bulk analysis of ventricular tissue from the K depleted animals (Ward and Cameron, 1984; Ward, 1988) which showed that there was no change in concentration in either K or Na. In rabbits cardiac tissue adapts to K depletion by an increase in Na pump density (Ward and Cameron, 1984; Ward, 1988). The increased number of cells with higher K/Na ratios in the preparation isolated from the K depleted animals probably reflects an increased ability of the cardiac myocytes to take up K from the isolation medium in which a normal level of K was maintained.

Summary and Outlook

Cryofixation and X-ray microanalysis have become established techniques for the study of diffusible elements in cells in tissue culture. The majority of studies have been on cells which have been grown as adherent monolayers. The results presented here for isolated cardiac myocytes show several features which are typical of isolated cells in suspension. When freshly isolated cells are studied non viable cells are present in the population, these can be identified both by their morphology and by their elemental content. Mean concentrations of K and Mg in the viable, rod-shaped, cells are comparable to values for these elements obtained from bulk analysis of cardiac tissue. The mean concentration of Na in the freshly isolated myocytes is increased compared to the value from analysis of tissue. The increase in concentration of Na is due to the presence of individual cells with higher Na concentrations. The investigations on myocytes from K depleted animals show that changes which occur in the transport properties of the membrane of the cardiac myocytes after experimental manipulation can be detected in the freshly isolated cells. The use of X-ray microanalysis of freshly isolated cells will allow us to gain information about elemental concentrations in normal tissues and any changes which occur after the onset of disease.

Acknowledgements

Work on isolated cardiac myocytes was supported by The British Heart Foundation, earlier work on thymocytes by The Sir Jules Thorn Charitable Trust. Funds from The Weston Foundation enabled purchase of a Reichert KF80 plunge freezing device. I would like to thank Dr. JPT Ward, Departments of Physiology and Medicine, U.M.D.S, for supplying the isolated cardiac myocytes and Dr. M Osborne, Dr. A Spencer and Mrs. L Tomkins, Department of Physiology, Medical School, University of Birmingham, for use of, and help with, their analytical microscope.

References

Abraham EH, Breslow JL, Epstein J, Chang-Sing P, Lechene C (1985) Preparation of individual human diploid fibroblasts and study of ion transport. Am J Physiol 248:C154-C164

Acker H, Pietruschka F, Zierold K (1985) Comparative measurements of potassium and chloride with ion-sensitive microelectrodes and X-ray microanalysis in cultured skeletal muscle fibres. In Vitro Cell Dev Biol 21:45-48

Buja LM, Hagler HK, Parsons D, Chein K, Reynolds RC, Willerson JT (1985) Alterations of ultrastructure and elemental composition in cultured neonatal rat cardiac myocytes after metabolic inhibition with iodoacetic acid. Lab Invest 53:397-412

Ceder O, Roomans GM, Hosli P (1982) Increased calcium content in cultured fibroblasts from trisomy patients: comparison with cystic fibrosis fibroblasts. Scanning Electron Microsc 1982/II:723-730

Dow JW, Harding NGL, Powell T (1981a) Isolated cardiac myocytes II Functional aspects of mature cells. Cardiovasc Res 15:549-579

Dow JW, Harding NGL, Powell T (1981b) Isolated cardiac myocytes I Preparation of adult myocytes and their homology with intact tissue. Cardiovasc Res 15:483-514

Gupta BL, Hall TA (1978) Electron microprobe X-ray analysis of calcium. Ann NY Acad Sci 307:28-51

Hook GR, Elin RJ, Hosseini JM, Swyt C, Fiori CE (1986) A sample preparation for quantitative determination of magnesium in individual lymphocytes by electron probe X-ray microanalysis. J Microsc 141:69-78

Jones RT, Johnson RT, Gupta BL, Hall TA (1979) The quantitative measurement of electrolyte elements in nuclei of maturing erythocytes of chick embryo using electron-probe X-ray microanalysis. J Cell Sci 35:67-85

James-Kracke MR, Sloane BF, Shuman H, Karp R Somlyo AP (1980) Electron probe analysis of cultured vascular smooth muscle. J Cell Physiol 103:313-322

Kirk RG, Andrews SB, Lee P (1983) An X-ray microanalysis study of cation changes during development in erythropoetic cells. Scanning Electron Microsc 1983/II:793-800

Lechene C (1986) Electron-probe analysis of cultured cells. An NY Acad Sci 483:270-283

LeFurgey A, Ingram P, Mandel LJ (1986) Heterogeneity of calcium compartmentation: electron probe analysis of renal tubules. J Memba Biol 94:191-196

Lew VL, Hockaday A, Sepulveda MI, Somlyo AP, Somlyo AV, Ortiz OE, Bookchin RM (1985) Compartmentalization of sickle-cell calcium in endocytotic inside-out vesicles. Nature 315:586-589

Osornio-Vargas AR, Berezesky IK Trump BF (1981) Progression of ion movements during acute myocardial infarction in the rat an X-ray microanalysis study. Scanning Electron Microsc 1981/II:463-472

Polimeni P (1974) Extracellular space and ionic distribution in rat ventricle. Am J Physiol 227:676-683

Poole-Wilson PA, Cameron IR (1975) Intracellular pH, and electrolytes in cardiac and skeletal muscle. Am J Physiol 229:1229-1304

Somlyo AP, Somlyo AV, Shuman H (1979) Electron probe analysis of vascular smooth muscle: composition of mitochondria, nuclei and cytoplasm. J Cell Biol 81:316-335

Spencer AJ, Haddon SJ, Osborne MP, Starkey WG, Collins J, Wallis TS, Clarke GJ, Worton KJ, Stephen J (1988). X-ray microanalysis of murine gut epithelia infected with EDIM rotavirus. Inst Phys Conf Ser 93:585-586

Tvedt KE, Halgunset J, Kopstad G, Haugen OA (1988) Quick sampling and perpendicular cryosectioning of cell monolayers for the X-ray microanalysis of diffusible elements. J Microsc 151:49-59

Von Zglinicki T, Bimmler M (1987) Intracellular water and ionic shifts during growth and ageing of rats. Mech Ageing Dev 38:179-187

Von Zglinicki T, Bimmler M, Purz HJ (1986) Fast cryofixation technique for X-ray microanalysis. J. Microsc 141:79-80

Ward JPT (1988) Tissue electrolytes in the rabbit following chronic dietary potassium depletion. Q J Exp Physiol 73:1005-1008

Ward JPT, Cameron IR (1984) Adaptation of the cardiac muscle sodium pump to chronic potassium deficiency. Cardiovasc Res 18:257-263

Warley A (1986) Elemental concentrations in isolated rat thymocytes prepared for cryofixation in the presence of different media. J Microsc 144:183-191

Warley A (1987) X-ray microanalysis of cells in suspension and the application of this technique to the study of the thymus gland. Scanning Microsc 1:1759-1770

Warley A, Kendall MD, Morris IW (1986) Problems associated with the preparation of cell suspensions for X-ray microanalysis highlighted by the comparison of results with those obtained from tissue sections. In: Muller M, Becker RP, Boyde A, Wolosewick JJ (eds) The science of biological specimen preparation for microscopy and microanalysis 1985. SEM, AMF O'Hare IL 60666 pp141-145

Wendt-Gallitelli MF, Wolburg H (1984) Rapid freezing, cryosectioning, and X-ray microanalysis on cardiac muscle preparations in defined functional states. J Electron Microsc Techn 1:151-174

Wroblewski J, Roomans GM (1984) X-ray microanalysis of single and cultured cells. Scanning Electron Microsc 1984/IV:1875-1882

Wroblewski J, Roomans GM, Madsen K, Friberg U (1983) X-ray microanalysis of cultured chondrocytes. Scanning Electron Microsc 1983/II:777-784

Wroblewski R, Edstrom L (1984) Recent applications of X-ray microanalysis in muscle pathology. Scanning Electron Microsc 1984/I:249-259

Zierold K (1981) X-ray microanalysis of tissue culture cells in SEM and STEM. Scanning Electron Microsc 1981/II:409-418

Zierold K, Schafer D (1988) Preparation of cultured and isolated cells for X-ray microanalysis. Scanning Microsc. (In Press)

Zierold K, Schafer D, Pietruschka F (1984) The element distribution in ultrathin cryosections of cultivated fibroblast cells. Histochemistry 80:333-337

Conference Discussion

K. Zierold: How is the K/Na ratio in the cells changed by recovery time after isolation?
A. Warley: The experiments for recovery after isolation have not yet been carried out. However, from experiences gained on the study of isolated thymocytes (and the experiences of other authors using different cell types) I would expect that K/Na would be increased after isolation.

Th. von Zglinicki: Is the K/Na ratio distribution in K-depleted animals in fact bimodal?
A. Warley: It is possible that this distribution is bimodal, the group with high K/Na ratios representing the cells in which membrane integrity has not been disturbed and which are able to maintain the high ratios because of increased Na pump density in their membranes.

B.L. Gupta: With reference to the large spread in the Na concentrations of your normal muscle cells, where in the cells did you make your measurements? Na may not be uniformly distributed in the cell e.g. between A and I band regions. Nuclei are capable of accumulating vast quantities of Na in moribund cells.
A. Warley: The measurements were made using a wide probe in the myofibrillar regions of cells; therefore, there may be some variation due to uneven distribution. As yet, I have not looked at the concentrations of elements in the nuclei of these cells.

A.J. Morgan: You can distinguish two subpopulations of cells 'rods' and 'rounds' in culture on the basis of morphological differences. Are there any differences in the rods that would permit you to subdivide these into further subpopulations on the basis of chemical composition? How do you propose to compare chemical heterogeneities between experimental groups?
A. Warley: It is clear that even on morphological considerations the rods are not a homogeneus population of cells and I

shall be doing more studies on this particular group of cells to see whether the group can be divided according to elemental concentrations. It is interesting that on the whole the rod cells from the K-depleted animals were more homogeneous and showed elevated concentrations of K, when compared to the cells from control animals. It may probably be better to undertake comparisons after a period of incubation which should 'settle' the elemental concentrations.

Ann LeFurgey: In myocytes from K-depleted animals are there changes in content of elements other than sodium and potassium e.g. calcium?

A. Warley: The studies on the intact myocardium have shown increases in calcium after animals are fed a K-repleted diet. Therefore, I would expect to see changes in this element in the isolated myocytes after a period of incubation.

X-Ray Microanalysis and Free Calcium Measurements in Cultured Neonatal Rat Ventricular Myocytes

Herbert K. Hagler, Alan C. Morris, and L. Maximilian Buja
Department of Pathology
University of Texas Southwestern Medical Center
5323 Harry Hines Blvd.
Dallas, TX 75235-9072
USA

INTRODUCTION

The response of cardiac myocytes to changes in calcium homeostasis is thought to be an important factor in cardiac myocyte injury under various conditions of ischemia, hypoxia and reperfusion injury(Allen and Orchard, 1983; Barry et al., 1987; Buja et al., 1985; Farber, 1982; Jennings and Reiner, 1981; Nayler, 1981). During the past several years advances in cryofixation, data collection and interpretation of electron beam excited x-ray microanalysis have been developed in our laboratory (Hagler et al., 1979; Hagler et al., 1983; Hagler and Buja, 1986). The use of x-ray microanalysis techniques have been important in understanding changes in total elemental content on a compartment by compartment and a cell by cell basis (Wendt-Gallitelli and Jacob, 1982; Wendt-Gallitelli and Wolburg, 1984; Wheeler-Clark and Tormey, 1987; Somlyo et al., 1985). A model of hypoxic and ischemic injury in our laboratory has been a primary culture of neonatal rat ventricular myocytes (Buja et al., 1985). Observations will be presented in this paper from this cultured cell model to make general points about biological x-ray microanalysis. The data analysis schemes which we have devised to look at a changing population of cells will be illustrated from this work. We will also present some data on intracellular free calcium measurements and their correlation with x-ray microanalysis measurements in this model.

EXPERIMENTAL MODEL

An experimental model system for use in x-ray microanalysis has some rather rigorous requirements in order to insure that the tissue is in a known physiological state prior to cryofixation, that the surface of the tissue is not mechanically damaged prior to cryofixation which would complicate the interpretation of the results and that the tissue is small enough to be amenable to cryofixation techniques. A system which satisfies the above requirements is the cultured neonatal rat ventricular myocyte model. For this model ventricular myocytes were isolated from the hearts of 2-3 day old rats and cultured using previously described methods(Buja et al., 1985). A calcium concentration of 1.8 mM was used. In order to manipulate the cells for freezing, they were placed on a collagen gel. The first addition of cells resulted in a light colonization of the gels. After three days the media with unattached cells was discarded and a second seeding of cells was added. After an additional 3-4 days the cells formed beating colonies of myocytes and were used for experiments.

CRYOFIXATION

The rapid cryofixation of tissue is necessary to preserve the elements of biological interest and to minimize the structural distortion of the tissue due to the formation of ice crystals. We have previously used a simple pair of pliers with copper jaws cooled to liquid nitrogen temperature for rapidly freezing of the tissue and cells(Hagler et al., 1983; Buja et al., 1985). In collaboration with the Gatan company, we developed an automated freezing device, a Cryosnapper Model 669 (Gatan, Pleasanton, CA), for more reproducible freezing conditions than was provided by the manual pliers. The cryosnapper is illustrated in Figure 1. It consists of a pneumatic cylinder which is operated by pressurized air, nitrogen or argon. Compressed gas is supplied to the device at 40 PSI. An internal regulator allows the adjustment of the driving pressure between 20-40 PSI. This regulated pressure controls the force of closure of the copper jaws. There are three positions which are controlled using solenoid valves, out, in, and release. The out position is used to cool the copper jaws in a bath of liquid nitrogen just prior to use. The jaws are then returned to the in position, Figure 1, which protects the freezing surfaces from contamination with frost. The cryosnapper is aimed using a small stainless steel wire and the out position selected, thus rapidly extending and closing the copper jaws onto both sides of the tissue or cells. The tissue is removed under liquid nitrogen by selecting the release position which allows the extraction of the tissue from the jaws. The advantages of this freezing device are reproducibility, adaptability for in vivo and in vitro experiments, and the capability of rapid freezing of the tissue without having to remount the tissue in a freezing head, thus permitting more precise control of the physiological state of the tissue or cells just prior to freezing.

Cultured neonatal myocytes have been frozen as previously described with the manual copper pliers(Buja et al., 1985), and, more recently, with the Cryosnapper. The cells in the culture plate were treated and harvested at appropriate experimental time points. Prior to the next step the copper jaws of the cryosnapper were precooled in a bath of liquid nitrogen. At a particular time point a 1 x 1 cm square of gel was scored with a scalpel blade and gently teased from the culture plate. The square of gel and cells were floated onto a small rectangle of filter paper and removed from the culture dish. The excess liquid on top of the cells was removed using gentle suction from below. When the liquid covering the cells was at a minimum, the cells were positioned under the aiming wire of the cryosnappers and the out position was selected. The jaws were returned to the liquid nitrogen bath and the release position selected so the cells and filter paper could be removed and placed into long term storage prior to cryosectioning.

CRYOSECTIONING

Cryosections were cut at -130°C at a thickness setting of 150 nm using a Dupont-Sorvall MT5000 ultramicrotome equipped with a FS1000 cryosectioning attachment. The sections were collected onto one side of a 50 mesh folding grid. The folding grid was previously coated with a formvar film and then coated with carbon to make the film conductive. The conductive film on the grid was necessary to insure reliable placement of the dry cut sections on the film well removed from adjacent grid bars. After section placement the grid was folded using a grid folding device illustrated in Figure 2. The grid folding device provides precision folding, thus the grid bars are in exact registration providing the maximum grid opening area to be maintained. The folded grid with frozen

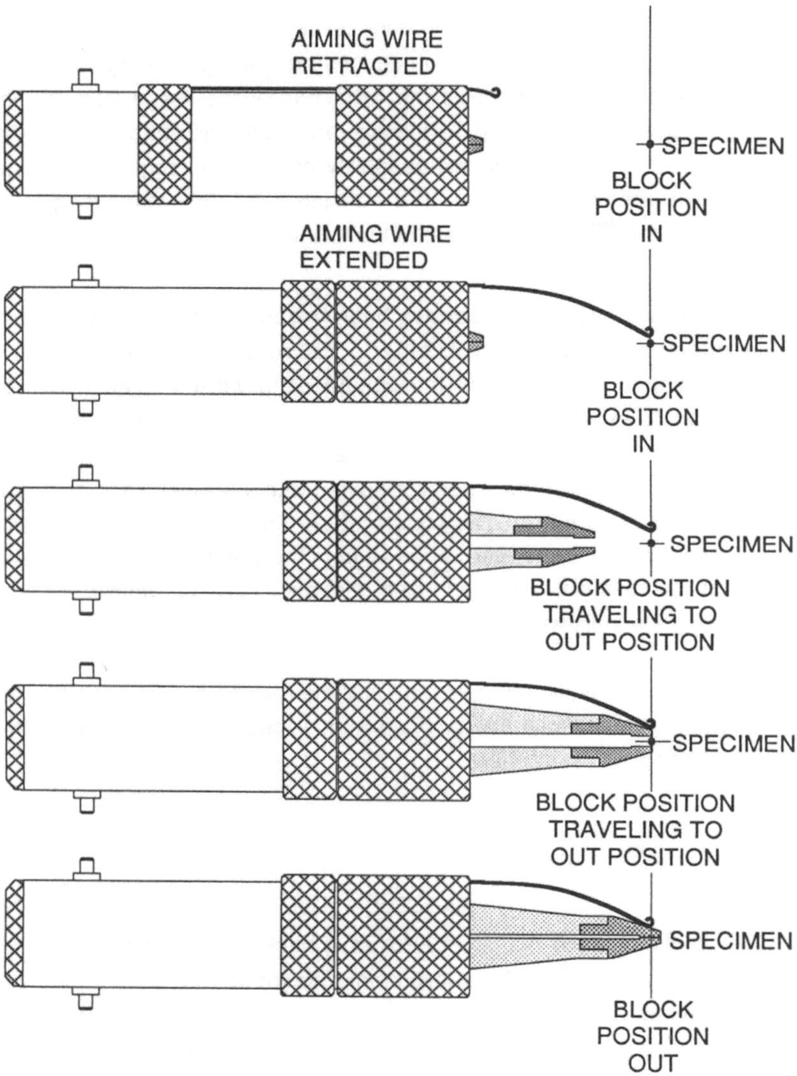

Figure 1 Illustration of the sequence of steps involved with aiming the cryosnapper, the path taken by the copper jaws during extension of the jaws. and the final closure of the jaws onto the tissue.

sections can be placed into a grid box and stored in a liquid nitrogen dewar until time to cryotransfer and collect x-ray data.

CRYOTRANSFER

The cryotransfer of frozen sections to the electron microscope has been used for a number of years (Zierold, 1982, 1984). The previously prepared sections on folded grids were transferred to a Gatan Model 626 cryotransfer stage(Hagler and Buja, 1986). The normal stage temperature during transfer to the electron microscope was -130°C. After observing the sections in the hydrated state, the stage temperature was set to -80°C and the sections freeze-dried in the microscope column. Drying judged to be complete after 20-30 minutes when the sections were able to sustain exposure to the electron beam for 100 seconds without visible mass loss in the rastered area. This term of no visible loss is not entirely correct as a mass loss of up to 25% could be present before it would become visible in the bright field STEM image. Should rapid mass loss occur at this step, additional drying time would be used until this condition was eliminated. Cryotransfer and drying in the electron microscope has the advantage of completely eliminating any concerns and questions regarding melting of lipids(Hagler and Buja, 1986) and/or rehydration which can occur when external freeze-drying procedures are used. Figure 3 illustrates some typical cryosections taken from control cultured neonatal rat ventricular myocytes, panel A and B, and after 60 minutes incubation in 1 mM ouabain, panel C. The cell swelling and calcium accumulation can be appreciated by comparing panel B and C.

X-RAY MICROANALYSIS PROCEDURES

Our initial x-ray data were obtained with a JEOL 100C (JEOL, Inc., Peabody, MA) electron microscope (Buja et al., 1985). More recently, a JEOL 1200EX electron microscope retrofitted with helium cryopumps has been used (Morris et al., 1989). The helium cryopumps provide an oil free vacuum system which are very efficient in removing water vapor. The vacuum achieved and routinely used during analysis was 3 x 10^{-7} Torr. A specially designed anticontaminator which operates at -170° C was also fitted into the objective lens. X-ray spectra were collected using a 30 mm^2 Si(Li) detector and a TN5500 x-ray analyzer(Tracor-Northern, Middleton, WC). The scanning transmission mode was used at 120 KV, 35° specimen tilt, and a beam current of 3 namps. Areas of interest within cells were analyzed in the selected area raster mode. A scan speed of 10 seconds per frame was used for all data collection. The x-ray spectra were collected at 10 eV/channel resolution from

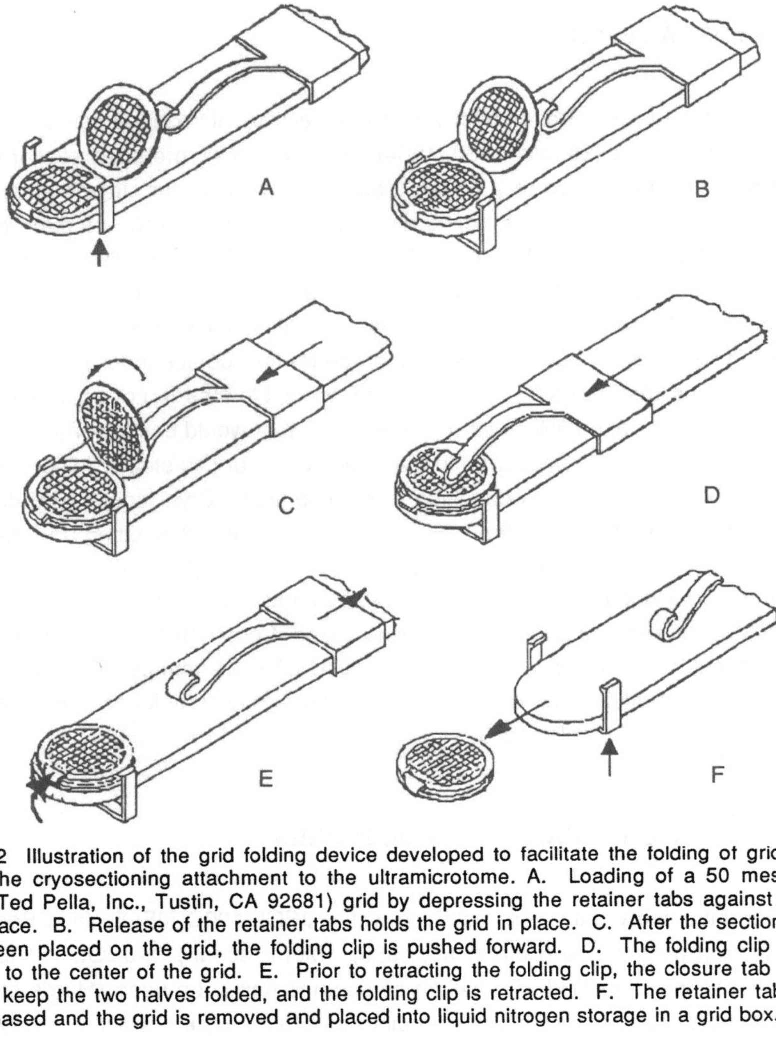

Figure 2 Illustration of the grid folding device developed to facilitate the folding of grids within the cryosectioning attachment to the ultramicrotome. A. Loading of a 50 mesh Pelco (Ted Pella, Inc., Tustin, CA 92681) grid by depressing the retainer tabs against a flat surface. B. Release of the retainer tabs holds the grid in place. C. After the sections have been placed on the grid, the folding clip is pushed forward. D. The folding clip is pushed to the center of the grid. E. Prior to retracting the folding clip, the closure tab is bent to keep the two halves folded, and the folding clip is retracted. F. The retainer tabs are released and the grid is removed and placed into liquid nitrogen storage in a grid box.

0 to 20 KeV for 100 seconds per measurement. Continuum was suppressed using a top hat filter, and peak intensities were obtained using multiple least squares fitting of a reference library of peak shapes to the unknown spectra. The relationship between peak-to-continuum ratio and elemental concentration was used to calculate concentration values(Hall et al., 1973). Analysis included using the first and second derivatives of the major (K alpha) potassium peak in the spectral analysis(Kitazawa et al., 1983). The continuum used was from 5.5 to 7.5

KeV. The continuum counts were corrected for the two support films and any contribution coming from the Cu grid prior to computing the peak-to-continuum ratios. The peak-to-continuum ratios were converted to concentrations by using weighting factors determined from standard curves obtained from freeze-dried sections of gelatin-glycerol specimens containing different concentrations of various elements(Hagler et al., 1983).

DATA ANALYSIS AND STATISTICS

Approaches to data analysis vary depending on the questions being addressed. In our case we would like to study very heterogeneous changes occurring in a population of cells over a period of time. A method has been developed which permits determination of a percentage of cells at a particular time point which are abnormal and describing those changes(Buja et al., 1985). The scheme developed consists of the following steps:

1. Three 100 second spectra were collected from cytoplasm, mitochondria, and whole cell for each cell identified in the thin sections. Cytoplasm was identified as an area of the intracellular space which is free of mitochondria, Z lines, and nuclear material. Mitochondria in the control tissue were identified as compact, electron dense ovoid structures. Cell measurements were collected from a large selected area raster of the cell excluding the nuclear region.

2. The cellular data from step 1 were collected for a large number of cells at each experimental time point. Data was collected from 40 to 100 cells from each time point.

3. The 3 spectra collected from the subcellular regions were averaged to get one mitochondria, one cytoplasm and one cellular measurement from each cell in the population sampled. It has been shown previously that the average of three 100 second measurements gives the same mean value as a single 300 second measurement. The 300 second measurement would have better counting statistics than the single measurement, hence less variability, but the mean value would be the same.

4. In order to separate the cells into normal and abnormal populations the control measurements were examined. Means and standard deviations were computed for each element of interest. It was assumed that even in the controls there would be a chance of having a few abnormal cells in this sample. Abnormal was defined as any measurement which was outside two standard deviations of the

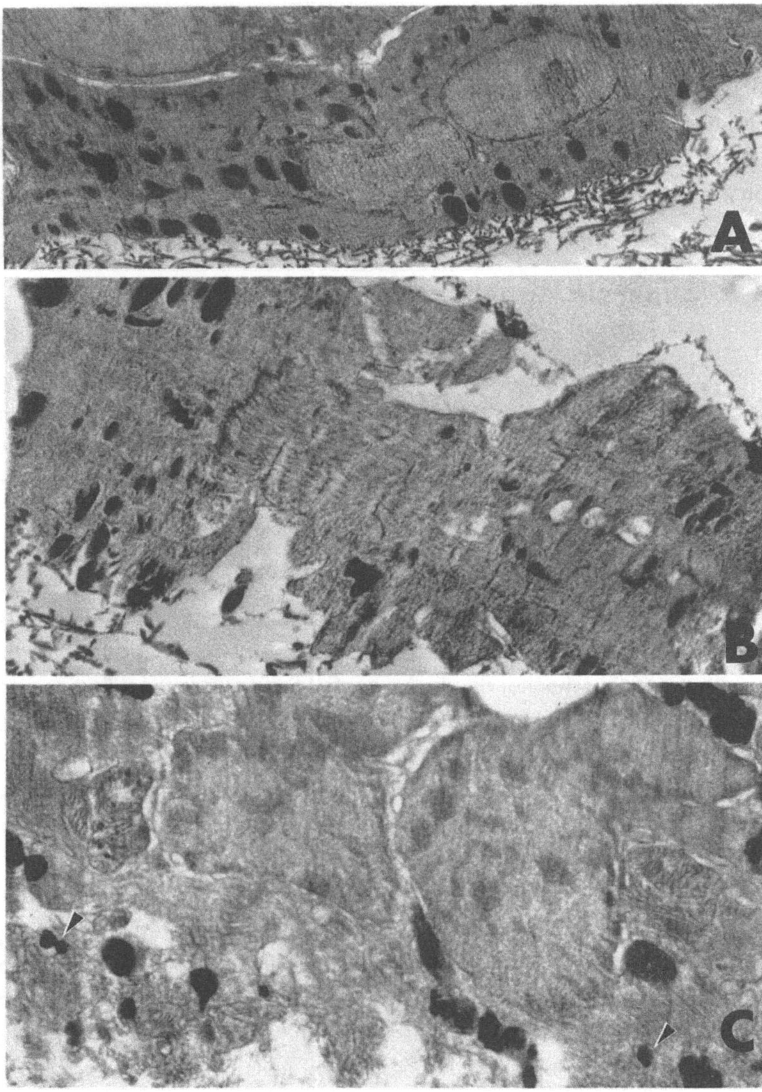

Figure 3 Scanning Transmission Electron Micrographs of Control neonatal rat ventricular myocytes A(8000x) and B(16,000x) showing normal cellular ultrastructure of cryofixed, cryosectioned and cryotransfered sections. C(17000x) Showing cellular injury after 2 hours of Na-K pump inhibition using 1 mM ouabain. This micrograph shows swollen mitochondria and areas of calcium accumulation (arrowheads) in nonmitochondrial locations.

mean of the control sample in the expected direction associated with injury. If abnormal cells were identified in the control sample they were labeled abnormal and a new control mean was computed. This mean and standard deviation were then used to identify cells at the experimental time points which were abnormal. In these experiments, injury was produced by depletion of ATP with the metabolic inhibitor, iodoacetic acid (IAA, 30 μM).

5. The grand means for each experimental group were calculated and the normal and abnormal means from each of the groups. This identifies the number of normal and abnormal cells over time as well as the percentage of cells which have become abnormal. Example data from neonatal rat ventricular myocytes are shown in Tables 1 and 2.

6. Analysis of variance and Duncan's multiple range test were applied to identify significant changes.

Table 1 Elemental Concentrations in Cytoplasm of Control and Iodoacetate(IAA)-Treated Cardiac Myocytes

Group	Na	Ca
Control(n=90)	30.2 ± 42.4^{A}	-1.7 ± 4.8^{A}
1 hr IAA(n=69)	67.5 ± 104.8^{B}	1.0 ± 13.8^{A}
1.5 hr IAA(n=102)	237.6 ± 136.4^{C}	12.0 ± 19.7^{B}
2 hr IAA(n=50)	255.5 ± 147.6^{C}	25.8 ± 43.0^{C}

Means (millimoles per kilogram dry weight) ± SD (n= number of spectra). In each vertical column, values with different superscripts are significantly different from each other by analysis of variance and Duncan's multiple range test.

Quantitative x-ray mapping is becoming more popular among many groups. The inappropriate use of mapping can make experimental changes very difficult to interpret. For most cell and tissue preparations a single section does not represent an unbiased sample. Mapping can be very useful if the section were an unbiased sample, or if some distribution were unknown and it was desired to determine specific morphological regions to later analyze in more detail using sampling techniques. The above described measurement scheme is a sampling technique which makes available many measurements from a large number of cells. As described above it was possible to have abnormal elemental distributions in control cells, which would only be identifiable using the statistical analysis described. In the case of muscle cells which are very anisotropic, a single section analyzed by x-ray mapping would constitute a biased sample of the population. Most morphometric studies and x-ray analytical studies can usually be performed with more efficiency and less sample bias by collecting a large number of samples. X-

Table 2 Normal and Abnormal Elemental Concentrations in Cytoplasm of Control and Iodoacetate(IAA)-Treated Cardiac Myocytes

Group	Na			Ca		
	Normal	Abnormal	Abnormal/total	Normal	Abnormal	Abnormal/total
Control	25±23	180.8	1/30	-1.9±2.5	4.4	1/30
1 hr IAA	36.6±29.5	214.3±150.2	4/23	-3.0±4.6	12.3±15.5	6/23
1.5 hr IAA	64.3±25.1	274.7±98.9	28/34	.5±1.8	14.9±19.6	27/34
2 hr IAA	102.3	257.7±128.3	15/16	.6±2.9	32±28.6	13/16

First two columns show means(millimoles per kilogram dry weight)±SD for normal and abnormal concentrations of the element, and the third column presents the ratio of the number of abnormal to total number of cells measured.

ray maps are in general collected under less than ideal conditions from biological thin sections. Most x-ray mapping has been done at room temperature because of the technical difficulty of maintaining a cold stage in a stable condition for the 2 to 24 hours of beam time necessary to complete the map. Although resulting problems with specimen stability and microscope stability may be compensated, they could cause registration problems in the data collected.

MEASUREMENT OF $[Ca^{+2}]_i$ TRANSIENTS

In contrast to measurement of total elemental concentration from subcellular regions by x-ray microanalysis, it has become feasible to measure the dynamic changes of ionized calcium in living cells(Allshire et al., 1987; Almers and Meher, 1985; Barry et al., 1987; Blinks, 1986; Buja et al., 1988; Grynkiewicz et al., 1985; Lee et al., 1987; Wier et al., 1987). For these measurements cells were cultured on laminin-coated round cover slips using method and loading conditions previously described(Morris et al., 1986, 1987, 1988a, 1988b, 1989). The coverslips were mounted in Sykes-Moore chamber and placed on a heated stage of a Nikon Diaphot microscope. Calcium transients from spontaneously contracting myocytes were measured using a Fluoroplex III 1000 microspectrofluorometer coupled to the epi-illuminator of the inverted microscope. Emission was measured at >470 nm by photon counting. The excitation wavelengths of 340 and 380 nm were provided using a ten sector chopper wheel which permitted 40 measurements per second to be collected. Background measurements were collected from a sample of cells not loaded with fura 2. Calibration of the fluorescence was performed in cell samples as recommended by Scanlon et al., 1987 and Peters et al., 1987 Rmax was determined using a non-fluorescent ionophore, 4-bromo-A23187 (20 μM) and Rmin was measured by treatment with 5 mM EGTA. The intracellular ionized calcium was calculated using the formula of Grynkiewicz et al., 1985 using a Kd of 225 nM. Figure 4 illustrates typical ionized calcium transients recorded from cells treated with the metabolic inhibitor, iodoacetate.

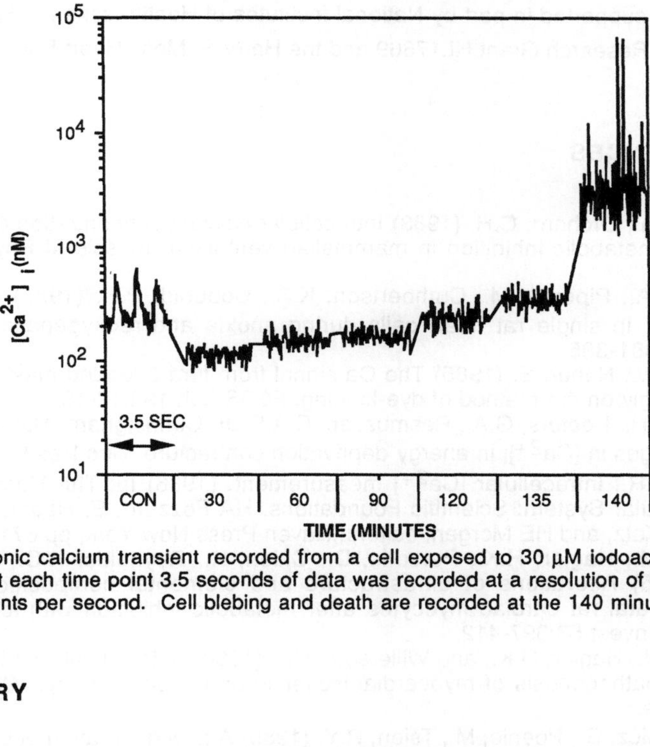

Figure 4 Ionic calcium transient recorded from a cell exposed to 30 μM iodoacetate for 140 minutes. At each time point 3.5 seconds of data was recorded at a resolution of 40 measurements per second. Cell blebing and death are recorded at the 140 minute time point.

SUMMARY

Some of the techniques and methods in routine use in our laboratory have been described in this presentation. The utility of x-ray microanalysis for some of our pathophysiological studies has been emphasized for the measurement of total elemental content of subcellular regions as well as for understanding the population dynamics with heterogenious pathophysiological changes. Recently the methods for using fluorescent indicators sensitive to ionized calcium have been added to these studies. It is quite useful to compare the results obtained with both techniques to develop a more complete picture of the elemental changes occuring during the time course of development of cell injury. The future is promising for using additional fluorescent indicators to correlate changes in pH, Mg, and Na with the x-ray analytical results.

ACKNOWLEDGEMENTS

The authors gratefully acknowledge the support of Mr. Dennis Bellotto for the cryosectioning and x-ray analysis and Ms. Donna Buja for the cultured cells. This work was supported in part by National Institutes of Health Ischemic Specialized Center of Research Grant HL17669 and the Harry S. Moss Heart Fund, Dallas, TX.

REFERENCES

Allen, D.G., Orchard, C.H. (1983) Intracellular calcium concentration during hypoxia and metabolic inhibition in mammalian ventricular muscle. J Physiol 339:107-122.

Allshire, A., Piper, H.M., Cuthbertson, K.S., Cobbold, P.H. (1987) Cytosolic free Ca^{2+} in single rat heart cells during anoxia and reoxygenation. Biochem J 244:381-385.

Almers, W., Neher, E. (1985) The Ca signal from fura-2 loaded mast cells depends strongly on the method of dye-loading. FEBS Lett 192:12-18.

Barry, W.H., Peeters, G.A., Rasmussen, C.A.F. Jr, Cunningham, M.J. (1987) Role of changes in $[Ca^{2+}]_i$ in energy deprivation contracture. Circ Res 61:726-734.

Blinks, J.R.: Intracellular $[Ca^{2+}]$ measurement. (1986) In: The Heart and Cardiovascular System: Scientific Foundations. HA Fozzard, E. Haber, RB Jennings, AM Katz, and HE Morgan, editors, Raven Press/New York, pp 671-702.

Buja, L.M., Hagler, H.K., Parsons, D., Chien, K., Reynolds, R.C., Willerson, J.T (1985) Alterations of ultrastructure and elemental composition in cultured neonatal rat cardiac myocytes after metabolic inhibition with iodoacetic acid. Lab Invest 53:397-412.

Buja, L.M., Hagler, H.K., and Willerson, J.T.. (1988) Altered calcium homeostasis in the pathogenesis of myocardial ischemic and hypoxic injury. Cell Calcium, in press.

Grynkiewicz, G., Poenie, M., Tsien, R.Y. (1985) A new generation of Ca^{2+} indicators with greatly improved fluorescence properties. J Biol Chem 260:3440-3450.

Hagler, H.K. and L.M. Buja. (1986) Effect of specimen preparation and section transfer techniques on the preservation of ultrastructure, lipids and elements in cryosections. J. Microsc. 141:311-317.

Hagler, H.K., Lopez, L.E., Flores, J.S., Lundswick, R. and Buja, L.M.. (1983) Standards for quantitative energy dispersive X-ray microanalysis of biological cryosections: validation and application to studies of myocardium. J. Microsc. 131:221-234.

Hagler, H.K. and Buja, L.M. (1986) Effect of specimen preparation and section transfer techniques on the preservation of ultrastructure, lipids and elements in cryosections. J. Microsc. 141:311-317.

Hall, T.A., Clarke-Anderson, H., and Appleton, T.. (1973) The use of thin specimens for x-ray microanalysis in biology. J. Microsc.(Oxford) 99:177-182.

Kitazawa, T., Shuman, H. and Somlyol, A.P. (1983) Quantitative electron probe analysis: problems and solutions. Ultramicroscopy 11:251-275.

Morris, A.C., Williams, P.K., Lattanzio, F.A., Bellotto, D.J., Hagler, H.K., Willerson, J.T., Buja LM (1986) Altered calcium homeostasis induced by Na^+, K^+ pump inhibition in cultured ventricular myocytes (abstract). Circulation 74(Suppl II):II-418.

Morris, A.C., Williams, P.K., Bellotto, D.J., Hagler, H.K., Willerson, J.T., Buja, L.M. (1987) Effects of Na^+, K^+ pump inhibition and metabolic inhibition on intracellular Ca concentration and localization in cultured neonatal rat cardiac myocytes (abstract). Fed Proc 46:1095.

Morris, A.C., Hagler, H.K., Buja, L.M. (1988a) Alterations in ionic calcium homeostasis in neonatal rat ventricular myocytes with metabolic inhibition (abstract). FASEB J 2:A1157.

Morris, A.C., Hagler, H.K., Nazeran, H., Buja, L.M. (1988b) Resolution of free calcium transients associated with contraction in cultured neonatal rat ventricular myocytes (abstract). Biophys J 53:601a.

Morris, A.C., Hagler, H.K., Willerson, J.T., and Buja, L.M. (1989) Relationship between calcium loading and impaired energy metabolism during Na^+, K^+ pump inhibition and metabolic inhibition in cultured neonatal rat cardiac myocytes. J.Clin.Invest., in press.

Nayler, W.G. (1981) The role of calcium in the ischemic myocardium. Am J Pathol 102:262-270.

Peeters, G.A., Hlady, V., Bridge, J.H.B., Barry, W.H (1987) Simultaneous measurement of calcium transients and motion in cultured heart cells. Am J Physiol 22:H1400-H1408.

Scanlon, M., Williams, D.A., Fay, F.S (1987) A Ca^{2+}- insensitive form of fura-2 associated with polymorphonuclear leukocytes: assessment and accurate Ca^{2+} measurements. J Biol Chem 262:6308-6312.

Somlyo AV, McClellan G, Gonzales-Serratos H, Somlyo AP (1985) Electron probe x-ray microanalysis of post-tetanic Ca^{2+} and Mg^{2+} movements across the sarcoplasmic reticulum in situ. J Biol Chem 260:6801-6807.

Wendt-Gallitelli, M.F. and Jacob, R. (1982) Rhythm-dependent role of different calcium stores in cardiac muscle: X-ray microanalysis. J Molec Cell Cardiol 14:87-492.

Wendt-Gallitelli, M.F., and Wolburg, H.. (1984) Rapid freezing, cryosectioning, and X-ray microanalysis on cardiac muscle preparations in defined functional states. J. Electron Microsc. Techn. 1:151-174.

Wheeler-Clark, E.S., Tormey, J. McD. (1987) Electron probe x-ray microanalysis of sarcolemma and junctional sarcoplasmic reticulum in rabbit papillary muscles: low sodium-induced calcium alterations. Circ Res 60:246-250.

Wier, W.G., Cannell, M.B., Berlin, J.R., Marban, E., Lederer, W.G. (1987) Cellular and subcellular heterogeneity of $[Ca^{2+}]_i$ in single heart cells revealed by fura-2. Science 235:325-327.

Zierold, K. (1982) Preparation of biological cryosections for analytical electron microscopy. Ultramicroscopy, 10:45-54.

Zierold, K. (1984) The morphology of ultrathin cryosections. Ultramicroscopy, 14:201-210.

Conference Discussion

R.L. Ornberg: In the discussion of freeze-drying or dehydrating paradigms you implied that once "all" of the water was removed (i.e. following the removal of water at T $\geq$ -40°C) you experienced no mass loss. Is this correct? We (and others, D. Johnson and K. Izutsu) report 25% mass loss in dry sections.
H.K. Hagler: We routinely remove the water from the cryosections using a stage temperature of -80°C, an anticontaminator at -170°C in a helium cryopumped JEOL 1200. Since there is no direct measurement when the water has been removed, we use a practical test of probing the specimen at high magnification 30,000 - 60,000x, 120 keV, and a 3 nA beam current. If we see noticable mass loss as visualized in brightfield STEM, then we prolong the drying time. "Dry" is defined as not being able to visualize a square raster pattern etched into the specimen when changing the magnification back to low magnification, 5000x. If the anticontaminator is not used, then the specimen tends to accumulate water from the microscope vacuum and cause mass loss. We had to redesign the standard anticontaminator because it was not efficient or cold enough to protect the specimen even at -80°C stage temperature. We undoubtedly have mass loss, but it is much less than the reported 25 %.

R. Reichelt: How do you explain the experimental finding that specimen areas pre-irradiated by electrons appear to be more resistant against air exposure than non-irradiated ones?
H.K. Hagler: My guess is that the electrons damage the proteins in some way to change them from a hydrophilic state to a hydrophobic state, thus, on exposure to room air, the irradiated areas do not absorb as much water as the non-irradiated portions of the sections. This is speculation on my part and I have no experimental data to confirm exactly what modifications are occurring.

R. Rick: In my experience there can be significant shrinkage of freeze-dried sections under the beam, if they do not properly adhere to the supporting film. In fact, we find it necessary to sandwich the sections between two films to obtain best stability.

H.K. Hagler: We have also observed that it is necessary to sandwich the sections between two formvar films for best stability of the sections. When the sections are well attached to the supporting film or films, then the sections are very stable. We encountered problems with the freeze-drying of sections between the two films until we learned the trick of poking holes in the periphery of the grid so the water vapor could excape from between the formvar films. Until we learned this, we had trouble obtaining "dry" sections.

J. Sommer: How do you recognize a frozen section to be attached to the film substrate on a grid as opposed to being loosely apposed to it?

H.K. Hagler: A section's attachment to the support film can be evaluated by using stereo pairs or rotating the specimen rod $\pm$ 20° while observing the image of the section. A better practical criteria is: if the section is stable during the analysis, then it is well attached. We routinely analyze only from sections that are stable under analytical microscope conditions.

K. Zierold: Your arguments to keep freeze-dried cryosections at a temperature below -80°C in order to avoid distortions by shrinkage of the section due to water sublimation and lipid melting at higher temperatures are conceivable. On the other hand, it is well known that not all water can be sublimed from the cell at -80°C (I. Wildhaber, H. Gross, H. Moor, 1982, J. Ultrastruct. Res. 80:367-373). Not completely dried sections containing (bound) water are very sensitive to radiation damage resulting in mass loss. How have you coped with this problem during X-ray analysis?

H.K. Hagler: We find that we can remove enough, maybe not all, of the water at -80°C to provide very stable sections. We also use our redesigned anticontaminator at -170°C because we found that the standard anticontaminator did not prevent deposition of ice on the specimen at -80°C and this contamination leads to severe mass loss on irradiation. See also answers to questions by R.L. Ornberg and R. Rick. We also routinely use a sandwich of support films, as suggested by R. Rick. Using these techniques and criteria we routinely obtain stable sections for X-ray analysis.

J. Morgan: You suggest that you routinely assess mass loss by visualizing your cryosections; the assumption being that if you cannot see it, then it does not occur at a level that, significantly affects your quantitation. But how much mass loss would you estimate that your specimen must suffer before you are able to "see it"? Secondly, do you consider that the stability of a given cellular compartment is representative of the matrix stability of all other compartments within a section that are destined for analysis?

H.K. Hagler: I would estimate that the specimen would have to suffer a mass loss of 15 - 25 % before we would be able to see it. We have also plotted the continuum (4.5 - 7.5 keV) as a function of time and found a linear relationship over the 100 seconds of analysis. This can be used to monitor any mass loss. The stability of a particular cellular compartment is not necessarily the same as other cellular compartments, for example, mitochondria have the best stability whereas cytoplasm (areas without discernable myofilaments) are the least stable. The assumption that cryostandards using a 20 % gelatin matrix behave the same as our sections under analytical conditions may not be correct when moving between different compartments.

Th. von Zglinicki: We found rather conflicting results on the transfer of frozen-dried cryosections through room air, even obtained on the same system (liver). Could you imagine reasons for these differences?

H.K. Hagler: Our original goal in looking at exposure of sections to room air was to identify worst case conditions. The worst case is exposure to air at a temperature near the dew point. This causes massive damage and redistribution of elements. If the exposure is at temperatures $>35^{o}C$, then structural and elemental changes are not noticable in the thicker section (150 - 200 nm) routinely used for X-ray analysis. If one looks at approximately 100 nm sections, structural changes can still be observed even at elevated temperatures. The elements which appear most sensitive to translocation appear to be Na and Cl. The real message is that external freeze-drying and exposure of dry sections to room air transfers must be done with great care, otherwise, this artefact can be introduced into your data. This also assumes that the lipid content of the tissue is very low so that lipid melting and redistribution are not going to cause problems.

1 μm Thick Frozen Hydrated/Dried Sections for Analysing Pericellular Environment in Transport Epithelia; New Results from Old Data

Brij L. Gupta
Department of Zoology
Cambridge University
Cambridge CB2 3EJ
England

Technical Preamble

During the period 1973-84, we carried out electron probe X-ray microanalysis (EPXMA) of 1/um thick frozen-hydrated sections on a variety of ion and water transporting epithelia. The instrumentation and methodology have been described in a number of previous publications (Gupta et al., 1977; Gupta and Hall, 1979, 1981b, 1982; Hall and Gupta, 1983). Hall (1986) has discussed the advantages of using ~1/um thick rather than ultrathin (~200 nm) cryosections for EPXMA, in spite of the limited structural image resolution afforded, especially in a scanning microscope operating at 45 - 50 keV (JEOL JXA-50A). However, most of our biological results published up to 1983 were obtained without the facility of a fully comprehensive X-ray data reduction and conversion into mM kg^{-1} wet or dry mass or mM l^{-2} H_2O, required by Hall's continuum normalization procedure, especially with respect to actual Z^2/A correction for every microvolume analyzed. Corrections for all the X-rays from extraneous sources may not have been as accurate because of limitations of computation facilities. The completion of the Link System QUANTEM/FLS soft-ware together with its dedicated computer and fast on-line print-out facilities (Gupta and Hall, 1982) allowed us to obtain more refined data on many tissues analyzed previously. The availability of a comprehensive standards file both for the chemical elements and the mass thickness (= section thickness under the beam) in the soft ware allowed us not to use bath Ringer + Dextran as peripheral standard in our in vitro preparations, as done previously; although we always established that 10 - 20 % Dextran (Mr 250,000) added to bath Ringers surrounding the epithelia in vitro did not measurably effect the transport function (Barnard et al., 1984). However, a Ringer's solution with 20 % Dextran frozen and cut with every tissue block was still required as a 'peripheral standard' in every section in order to establish the exact level of hydration of the sections. In the studies reported here, this 'peripheral standard' was added on the Ringer-surrounded piece of tissue (on the serosal side in ileum, Table 1) immediately before rapid freezing a la Münich group (Rick et al., 1982). Local dry mass and hence water fraction for every analyzed field were obtained by using formulations involving mas fractions of all the nuclides analyzed in every microvolume (Hall and Gupta, 1982; Table 2). Great care was taken that local continuum counts from the sections after drying were collected from an area of the subcompartment not previously analyzed in a frozen-hydrated state and hence had not suffered the latent mass loss due to beam damage. Here I report results concerning the ion and water composition of various pericellular compartments which in the normal living tissues always contain a 'matrix' of mucoid substances constituting a local dry mass of some 5 % to 30 %, depending on the site and the physiological state of the transport function, even though such pericellular compartments usually appear 'empty' in conventionally prepared material for light and electron microscopic histology without special precautions to fix mucoids. A more comprehensive discussion of the subject has recently been published (Gupta, 1989).

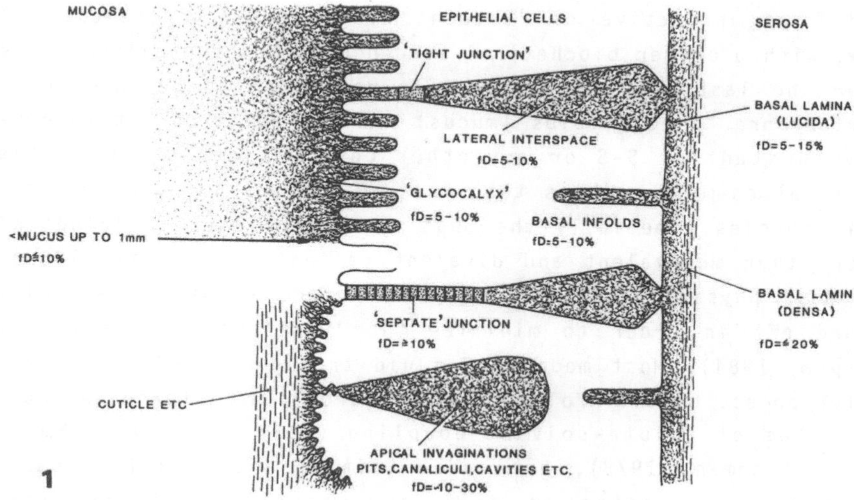

1

<u>Pericellular compartments.</u> Fig. 1 is a highly schematic summary of the pericellular compartments in ion and water transporting epithelia. The top half is typically representative of vertebrates and the bottom half of invertebrates, although features such as apical invaginated cavities can be found in either of the groups. The presence of polyanionic mucoid matrices at these sites has been established by using the chemicals such as alcian bue GX, Ruthernium red, lanthanates etc., especially in mammalian tissue (Oschman, 1978). The local dry mass concentrations (f_D) of the matrices estimated in our studies of frozen-hydrated sections are given for each site. The biochemical composition has been extensively analyzed for the mucosal mucus and the serosal materials (basal laminae) both in vertebrates (Fransson, 1987) and invertebrates (Denny, 1983). The major components are highly acidic glycoproteins, proteoglycans, and glycosaminoglycans with up to 95 % carbohydrate residues rich in COO^-, PO_3^-, SO_3^-. Highly sulphated heparan sulphates (up to 4 sulphates per tetrasaccharide) are abundant in glycocalyces and basal laminae and are likely to be secreted into lateral interspaces and other invaginations of the membranes. All these sites often showed high S and some P contents in all our analytical measurements. Previously, we considered this S

and P as indicative of 'unseen' cytoplasmic components, but now, with a better biochemical evidence from the published work over the last ten years, there is no reason to believe this. Furthermore, free mucoids (mucus) are now believed not to have any substantial S-S or any other covalent cross-links. The physical chemistry shows that the -ve charges in such polyanionic mucoids especially the SO_3^- residues prefer to 'bind' K^+ over other monovalent and divalent cations (see Scott, 1989). In most physiological studies the surface mucus is carefully wiped off in order to minimize the 'unstirred' layer effect (Gupta, 1984). Most models of fluid transport include lateral interspaces, basal infolds, apical microvillar interspaces etc. as sites of solute-solvent coupling (Berridge and Oschman, 1972; Oschman, 1978), but model them as 'free fluid' spaces with no ion-binding or diffusion-restricting properties. However, in recent years, evidence has accumulated that the surface mucus buffers H^+, K^+, Na^+, and Cl^- activities and maintains a distinct 'microenvironment' at the epithelial surfaces (Engelhardt et al., 1986; Kirsch, 1987). The data presented here and elsewhere in recent literature show that similar considerations might apply to other sites.

Rabbit ileum. Fig. 2a represents a low magnification STEM image of a frozen-dried cryosection of the tissue we analyzed. The flat-sheet-preparation was physiologically monitored in vitro in an Ussing-type chamber for 30 minutes and then quick-frozen (Gupta et al., 1978). There was a 100 - 200 μm thick layer of mucus at the mucosal surface. The mammalian ileum is a fluid absorbing and secreting epithelium with "transport tone" (Berridge, 1984). Absorption occurs in the top 1/3 of the villus while secretion is localized in the crypt area. Fig. 3 provides an example of the distribution of Na, K, and Cl in the mucus filling the intervillar space under conditions of a net isotonic absorption. The ionic grdients are very similar to those reported for the mucosal mucus in the middle intestine of eel adapted to sea-water (Simonneux et al., 1987a,b). These workers used ion-selective microelectrodes and measured ionic

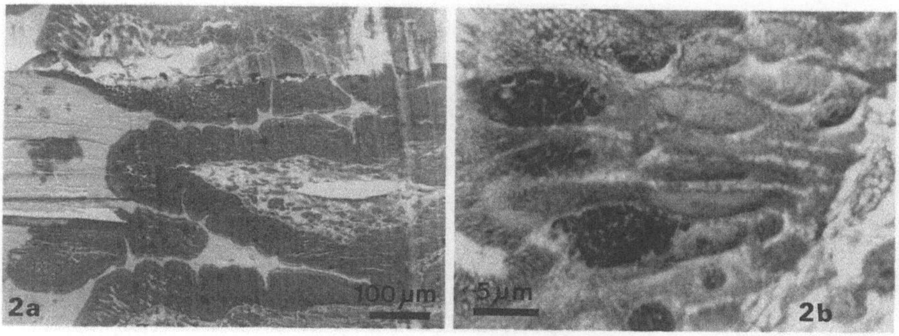

Fig. 2: STEM images of frozen-dried cryosections of rabbit ileum from in vitro samples: (a) villi, (b) goblet (mucous) cells.

activities. A gradient of pH along the intestinal villus in rat jejunum has also been reported (Daniel et al., 1985). We were unable to pursue this study further and the data available are very limited.

Table 1 shows the distribution of K ions near the apical surface of the enterocytes with highly dilated lateral interspaces (LIS) indicative of active fluid absorption. We previously reported (Gupta et al., 1978) that with 15 mM K in the bath Ringer there was 40 mM of K at the cell surface, supporting the observation of Zeuthen and Monge (1975) using ion-selective microelectrodes in vivo. This K concentration changes in parallel with the K concentration in the mucosal but not the serosal Ringers. However, even with Zero-K in the mucosal and serosal Ringers there was 5 $\pm$ 1 mM of K near the glycocalyx. This observation is consistent with the measurements of K activities at the mucosal surface of the colonic epithelium in guinea-pig (Engelhardt et al., 1986) and of the intestinal epithelium in fish (Simonneux et al., 1987b). We also found a correspondingly high K concentration in the basal lamina in the same preparations (not shown). At both these sites a higher activity of K^+ probably represents a recycling of cell K^+ with the polyanionic proteoglycans providing the key function of retaining K^+ ions leaking out of the cells (Gupta, 1989;

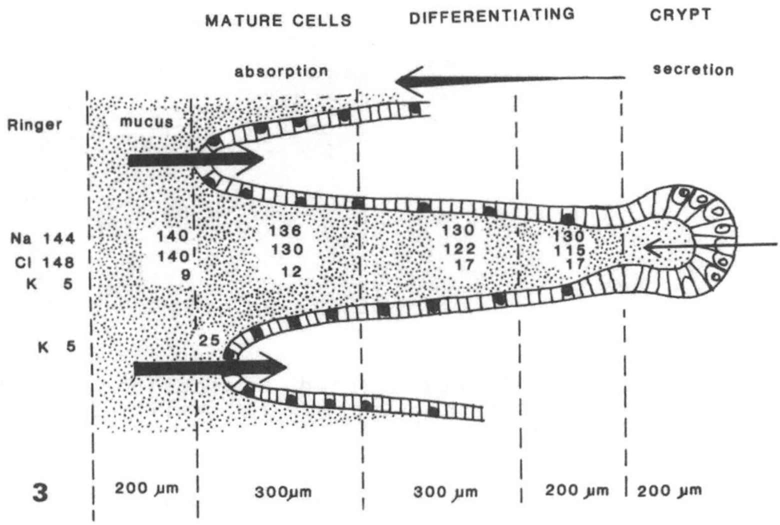

3

Kirsch, 1987). Table 1 further shows high concentrations of S and Ca at the mucosal surface.

By comparison the matrix in LIS, believed to be the primary sites of solute-solvent coupling (Spring and Hope, 1979; Gupta, 1984) has about 10 % dry weight: 14 mM of S and 6 mM of P. The latter probably constitutes fixed negative charges in the poly-anionic mucoid (glycocalyx, putative free proteoglycans). Apart from a high concentration of total Na and Cl, there are 12 mM of K and 5 mM of Ca. Both K and Ca are also considerably higher than in the bath Ringers. This suggests that most of the K and Ca are likely to be bound to polyanions. It is, therefore, not simple to realistically convert the total concentrations of ions measured with the microprobe into mM l^{-1} H_2O, because one does not know what the activity coefficients for dissociation are going to be (Gupta and Hall, 1979; Oschman, 1978; Gupta, 1989). The conversion into effective osmotic pressures poses a far greater porblem, because it additionally involves exact boundary conditions which are unknown. Polyanions also behave as gels which bind water and generate hydrostatic pressure, which in turn are directly influenced by their ionic environment.

Table 1: Elemental composition of apical mucus and lateral interspaces in rabbit ileum *in vitro* measured with EPXMA in frozen-hydrated/dried sections.

	Mucosal* Ringer	Apical Mucus	Lateral Interspace	Serosal* Ringer	Serosal** Standard
% Dry wt	-	12+2%	10+2%	<1%	20%
Na	144	138±8	164±10	144	144+4
Cl	148	140±3	162±3	148	148±3
K	5	25±5	12±3	5	5±1
Others	27	?	10?	27	27
S	-	51+11	14+4	-	nd
P	-	8±3	6±3	-	-
Ca	1	6±2	5±2	1	1+2
Mg	1	2±3	2±3	1	1±2
K	0	5+1	11+3	0	
K	15	39±4	18±3	15	

* not measured in the EPXMA
** Ringers + 20 % Dextran *Mr* 250 K (w/w) added before quick freez

---

Table 2: Dry weight fractions ($\%f_D$) of mucus granules in ileal goblet cells (see text).

	P	S	Cl	Na	K	Ca	Mg	n
mM kg^{-1} wet wt	30±3	212±11	30±2	18±8	36±6	60±6	58±14	21
mM kg^{-1} dry wt (short)	75±9	568±19	82±10	45±9	95±8	166±8	141±6	21
mM kg^{-1} dry wt (overnight)	92±12	703±23	99±11	59±11	125±9	210±10	178±12	8
short drying f_D %	40.0	37.3	36.8	39.4	37.7	36.1	41.1	21
long drying	32.6	30.2	30.3	30.5	28.8	28.6	32.5	8

Mean ± SD from all 7 nuclides: $\% f_D$ (short drying) = 38.3±1.8;

$\% f_D$ (long drying) = 30.5±1.6

Goblet cells. Recent work with quick-freezing-substitution has shown that mucus secreting 'goblet' cells synthesize, store, and secrete their product in much the same way as the well-known exocrine pancreatic cells (Sandoz et al., 1985; Verdugo et al., 1989). The cells face the problem of concentrating highly sulphated and sialated glycosaminoglycan polyanions into secretion granules, which would require extensive cross-links with dications (Mathews, 1970; Obrink et al., 1975). Conversely, ionic changes would be involved in exocytosis (Pollard et al.,1984; Green, 1987) and in the rehydration of mucus after release (Verdugo et al., 1989). Fig. 4 summarizes our results on the elemental concentrations in different synthetic compartments of the goblet cells (Fig. 2b) in rabbit ileum. Mature mucus granules acquire a dry mass concentration of nearly 40 % and consequently, have the highest electron density in our STEM image. During maturation, concentrations of Ca and Mg rise in parallel with the increase in dry mass and S while those of K drop. Changes in monovalent ions (K, Na, Cl) most likely reflect a mechanism for a solute-linked efflux of H_2O presumably driven by a H^+ pump, as postulated for some other cell systems (see Gupta, 1989). The results for Ca concentration are similar to those reported previously (Gupta and Hall, 1978).

At the exocytotic sites there is a decline in the concentration of S, Ca, and Mg leading to rehydration, as indicated by a drop in dry mass to about 21 %. A very high concentration of Cl supports the suggestion of Pollard et al. (1984) for a key role of this ânion in early exocytosis. The free mucus at the surface now has very low Ca and Mg, high Na and Cl, but also 15 mM of K (cf. Figs. 3, 4, and Table 1). It is known that at neutral pH, sulphates in polyanions readily exchange Ca^{2+} and Mg^{2+} with Na^+ resulting in rehydration and expansion (Mathew, 1970).

Anomalies in dry weight estimates. In frozen-hydrated sections dried in our routine for 5 - 15 minutes at about -100°C and 5×10^{-6} Torr in SEM column, the mucus granules had a dry mass

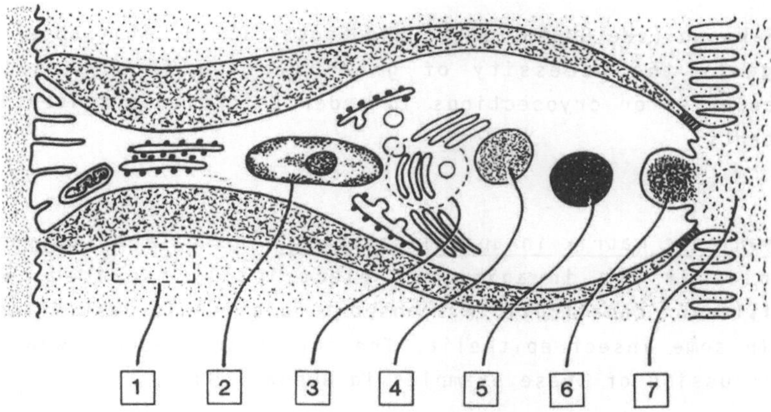

(mM kg^{-1}) wet weight)

	1	2	3	4	5	6	7
P	146	145	118	85	30	29	9
S	50	56	61	155	212	148	84
Cl	19	25	40	62	30	140	93
Na	15	10	49	44	18	89	129
K	120	136	112	90	36	39	15
Ca	2	3	6	32	60	39	6
Mg	5	6	16	22	58	39	2
(n)	(18)	(18)	(18)	(10)	(21)	(10)	(21)
$\%f_D$	25	15	24	30	38	21	13

Fig. 4: Mean elemental concentrations in different subcellular compartments in ileal 'goblet' cells: SD's were $\leq$ 10%.

($\%f_D$) of 38.3 $\pm$ 1.8 (Table 2). Some of the sections had to be left in the SEM overnight (circumstantial necessity) when the temperature of the cold-stage reached about 20°C. In such sections the mucus granules had a dry mass of 30.5 $\pm$ 1.6. There was no evidence of any selective loss of volatile elements (S, Cl). A substantially lower dry mass after long drying would, therefore, reflect a complete drying including the loss of bound-water, as also noted before by Gross (1987). A strong water-structuring ability of polyanionic proteoglycans is well-known (see Gupta, 1989, and Scott, 1989). This observation suggests the necessity of great caution in freeze-drying procedures for cryosections in order to estimate solvent water contents.

Polyanionic matrix in apical cavities. The role of polyanionic mucoids in the invaginated, electrically isolated, apical cavities or canaliculi in transport function is best illustrated in some insect epithelia. The interested readers would find a discussion of these examples in Gupta (1989).

Conclusions

The pericellular spaces in ion and water transporting epithelia contain polyanionic proteoglycans (mucoids) at concentrations varying from 5 to 30 % dry weight. Although the role of apically secreted (mucosal) mucus has been recognized as a protective covering (Allen, 1983; Flemström, 1987), attention has more recently been focused on the selective ion binding (Scott, 1989) and water structuring properties of polyanionic mucoids. These properties make these polymers behave as dynamic state Donnan systems with the ability of buffering an ionic microenvironment at the cell surfaces, which is distinct from the bulk extracellular fluids and the cytosol. A preference to bind K^+ and Ca^{2+}, especially by heparan and other sulphates (see Gupta, 1989), may be critical in conserving these ions at the epithelial cell surfaces. The EPXMA data presented here fully support the concept of conserving and recycling K^+ leaking out of the cell both at the apical and basolateral surfaces, thus reducing the energy load of transport on the cells. The physiologists must, therefore, take this into account and should model transporting epithelia as multicompartment systems (Gilles-Baillien, 1983; Gupta, 1984, 1989; Kirsch, 1987; Sackin and Boulpaep, 1975).

In the evolution of cellular life polysaccharides must have evolved early before the RNA and DNA. Scott (1989) has proposed that these polyanionic polymers might have been the earliest intracellular Donnan systems, and took up this function extracellularly after the evolution of proteins. The presence of K^+-binding heparan sulphates in the nucleus (Fedarko and Conrad, 1986) is perhaps vestigial and can explain why nuclei in general have a higher concentration of total K^+ in most cells examined with EPXMA (Gupta and Hall, 1982).

Acknowledgements

I am most grateful to Mr John Rodford, Mr Mike Day and Ms Francis Pang for graphic work and Mrs Joy Schreiber and Mrs Anne Philips for the typing.

References

Allen A (1983) Mucus - a protective secretion of complexity. Trends Biochem Sci 8:169-173

Barnard T, Gupta BL, Hall TA (1984) Effects of high-molecular-weight cryoprotectant Dextran on fluid secretion by *Calliphora* salivary glands. Cryobiology 21:559-569

Berridge MJ (1984) A general survey of the mechanism and control of intestinal fluid transport. In: Basic Science in Gastroenterology: Physiology of the Gut. JM Palade, SR Bloom, NA Wright, and AG Boulter (eds), pp. 225-246, Glaxo, Ware

Berridge MJ, Oschman JL (1972) Transporting Epithelia. Academic Press, New York-London

Daniel H, Neugebauer B, Kratz A, Rehner G (1985) Localization of acid micro-climate along intestinal villi of rat jejunum. Am J Physiol 248:6293-6298

Denny M (1983) Molecular biomechanics of molluscan mucus secretion. Mollusca 1:431-465

Engelhardt WV, Kuck U, Krause M (1986) Potassium microclimate at the mucosal surface of the proximal and the distal colon of guinea pig. Pflügers Arch 407:625-631

Fedarko NS, Conrad HE (1986) A unique heparan sulfate in the nuclei of hepatocytes: structural changes with the growth state of the cells. J Cell Biol 102:587-599

Flemstrom G (1987) Mucus and mucosal protection. In: Comparative Physiology of Environmental Adaptations, vol 1, R Kirsch, B Lahlou (eds), pp. 167-177. Karger, Basel

Fransson L-A (1987) Structure and function of cell-associated proteoglycans. Trends Biochem Sci 12:406-411
Gilles-Baillien M (1983) Several compartments involved in intestinal transport. In: Intestinal Transport. M Gilles-Baillien, R Gilles (eds), pp. 103-119, Springer-Verlag, Berlin
Green DPL (1987) Granule swelling and membrane fusion in exocytosis. J Cell Sci 88:547-549
Gross H (1987) High resolution metal replication of freeze-dried specimens. In: Cryotechniques in Biological Electron Microscopy. RA Steinbrecht, K. Zierold (eds), pp. 205-215, Springer-Verlag, Berlin
Gupta BL (1984) Models of salt and water flow across epithelia: an evaluation by electron probe X-ray microanalysis. In: Osmoregulation in Estuarine and Marine Animals. A Pequeux, R Coiles, L Bolis (eds), pp. 191-121. Springer-Verlag, Berlin
Gupta BL (1989) The relationship of mucoid substances and ion and water transport. Symp Soc Expt Biol 43, in press
Gupta BL, Hall TA (1978) Electron microprobe X-ray analysis of calcium. Ann NY Acad Sci 307:28-51
Gupta BL, Hall TA (1979) Quantitative electron probe X-ray microanalysis of electrolyte elements within epithelial tissue compartments. Fed Proc 38:144-153
Gupta BL, Hall TA (1981a) Microprobe analysis of fluid-transporting epithelia: evidence for local osmosis and solute recycling. In: Water Transport Across Epithelia. HH Ussing, N Bindslev, NA Lassen, O Sten-Knudsen (eds), pp. 17-35. Munksgaard, Copenhagen
Gupta BL, Hall TA (1981b) The X-ray microanalysis of frozen-hydrated sections in scanning electron microscopy: an evaluation. Tissue Cell 13:623-643
Gupta BL, Hall TA (1982) Electron Probe X-ray Microanalysis. Publication P128 in the series Techniques in Cellular Physiology. PF Baker (ed), pp. 1-52. Elsevier/North Holland, Amsterdam
Gupta BL, Hall TA, Moreton RB (1977) Electron probe X-ray microanalysis. In: Transport of Ions and Water in Animals. BL Gupta et el. (eds), pp. 84-143. Academic Press, London-New York-Sydney
Gupta BL, Hall TA, Naftalin RB (1978) Microprobe measurements of Na, K, and Cl concentration profiles in epithelial cells and intracellular spaces of rabbit ileum. Nature (London) 272:70-73
Hall TA (1986) Properties of frozen sections relevant to quantitative microanalysis. J Microsc 141:319-328
Hall TA, Gupta BL (1983) The localization and assay of chemical elements by microprobe methods. Q Rev Biophys 16:279-339
Kirsch R (1987) Mucus and epithelial ion exchange. In: Comparative Physiology of Environmental Adaptations, vol. 1. R Kirsch, B Lahlou (eds), pp. 158-168. Karger, Basel
Mathews MB (1970) Binding of calcium by proteoglycan of chondroitin sulphate. In: Chemistry and Molecular Biology of Intercellular Matrix. EA Balazs (ed), pp. 1121-1123. Academic Press, New York

Obrink BH, Pertoft H, Iverius P-H, Laurent TC (1975) The effect of calcium on the macromolecular properties of heparan sulphate. Conn Tissue Res 3:187-193
Oschman JL (1978) Morphological correlates of transport. In: Membrane Transport in Biology, vol. 3, Transport Across Multimembrane Systems. G Giebisch, DC Tosteson, HH Ussing (eds), pp. 55-93. Springer-Verlag, Berlin
Pollard HB, Pazoles CJ, Creutz CE, Scott JH, Zinder O, Hotchkiss A (1984) An osmotic mechanism for exocytosis from dissociated chromaffin cells. J Biol Chem 259:1114-1121
Rick R, Dörge A, Thurau K (1982) Quantitative analysis of electrolytes in frozen-dried sections. J Microsc 125:239-247
Sackin H, Boulpaep EL (1975) Models for coupling of salt and water transport. J Gen Physiol 66:671-733
Sandoz D, Nicolas G, Laine MC (1985) Two mucus cell-type revisited after quick-freezing and cryosubstitution. Biol Cell 54:79-89
Scott JE (1989) Ion-binding patterns of affinity depending on types of acid groups. Symp Soc Exp Biol 43: in press
Simonneaux V, Barra JA, Humbert W, Kirsch R (1987a) The role of mucus in ion absorption by the oesophagus of the sea-water eel (Anguilla anguilla L). J Comp Physiol B 157:187-199
Simonneaux V, Humbert W, Kirsch R (1987b) Mucus and intestinal exchanges in the sea-water adapted eel, Anguilla anguilla L. J Comp Physiol B 157:295-306
Spring KR, Hope A (1979) Fluid transport and the dimensions of cells and interspaces of living Necturus gallbladder. J Gen Physiol 73:287-305
Verdugo P, Aitken ML, Villalon M, Langley L (1989) Molecular mechanisms of product release and conditioning in cervical and respiratory goblet cells: role of extracellular polyions. Symp Soc Exp Biol 43: in press
Zeuthen T, Monge C (1975) Intra and extracellular gradients of electrical potential and ion activities of the epithelial cells of the rabbit ileum in vivo recorded with microelectrodes. Phil Trans Roy Soc London B71:277-281

Conference Discussion

J. Morgan: When gastric mucus layers develop local discontinuities, the underlying epithelial cells can experience severe degenerative events *in vivo*. Macromolecules clearly exert important regulatory influences on the extracellular environment of a cell. How much weight do you, therefore, attach to physiological observations on epithelial layers where surface mucus has been systematically removed?
B.L. Gupta: I am very skeptical of the results from such denuded epithelia studied *in vitro*. These preparations are designed to clean-up noise from the experimental data, but how relevant are these results for the actual function *in vivo*? I have previously discussed these issues in a review published in 1984. This consideration is of vital importance in using *in vitro* preparations for evaluating drugs for clinical application. This issue was debated in an SEB symposium on mucus in July 1988 (Symposium Soc. Exptl. Biol. 43, in press).

Th. von Zglinicki: Did you find changes in chemical composition of secretory granules and extracellular mucus in different states of secretory and/or absorptive stimulation?
B.L. Gupta: The data I presented were from glucose-stimulated isotonic absorption conditions. In unstimulated controls there was no significant difference in the gradients of K but Na and Cl gradients seemed to be less steep. However, not enough data are available. Mucus cells were the same under both conditions. We did look at the tissue under theophaline and cAMP stimulation as well as under artificially induced transepithelial osmotic gradients. There are piles of data with anomalies due to problems with QUANTEM-FLS in the development period. I have not "deconvoluted" these data so far, nor have I sorted out the data from the crypt area.

Distribution of Ions and Water in Epithelial Cells and Tissues

Roger Rick and Walter Schratt
Department of Physiology and Biophysics
Nephrology Research and Training Center
University of Alabama at Birmingham
Birmingham, Ala. 35294
U.S.A.

In tight epithelia, such as the toad urinary bladder and the mammalian collecting duct, a brisk transepithelial water flow can be induced by establishing a transepithelial osmotic gradient in the presence of antidiuretic hormone (ADH). Typically, water flow is accompanied by extensive cellular swelling (Peachey and Rasmussen,1961; Ganote et al., 1968; Kirk et al., 1984) which, unlike swelling which is not caused by transcellular water flow, is largely confined to the apical pole of the epithelial cell while the basal pole is only marginally distended (DiBona et al., 1969). Based on this finding, DiBona (1981) proposed that cytoplasmic water flow induces an intracellular standing gradient of solutes and, therefore, gradient of water activity.

We recently tested this hypothesis in both toad urinary bladder and rabbit cortical collecting duct by employing the technique of electron microprobe analysis (Rick and DiBona, 1987; Schratt et al., 1988). This method appears to be particularly well suited for resolving this issue because it allows direct examination of the solute composition of epithelial cells, even down to a subcellular level (Rick et al., 1982). In the presence of a transepithelial osmotic gradient and ADH, in both epithelial tissues a marked reduction in the intracellular ion concentrations of the apical cytoplasm was detectable, giving rise to an intracellular ion concentration gradient. No concentration differences were detectable in the absence of water flow. The results imply that the cytoplasmic pathway offers a significant hydraulic resistance and, thus, should be considered as an additional barrier to the transepithelial flow of water.

Materials and Methods

Toad urinary bladder. Large female toads of the species *Bufo marinus* were used. The animals were kept in dry plastic troughs and bathed in tap water for about 20 min on a daily basis. Toads were doubly pithed and each dissected urinary hemibladder was mounted over a plastic ring. The apical surface was jet-washed with amphibian Ringer's solution to remove the mucous layer. After initial equilibration with isotonic Ringer's solution (220 mOsm), the incubation was continued according to the experimental regime described below. Normal Ringer's solution contained (in mM): 110 NaCl, 2.5 $KHCO_3$, and 1 $CaCl_2$. Solutions of reduced osmolarity were obtained by lowering the NaCl concentration. All solutions were bubbled with air and had a pH ranging between 8.1 and 8.3. ADH (Pitressin, Parke Davis) was added to the serosal bath at an activity of 200 mU/ml.

After incubation, the mucosal surface of the epithelium was coated with a thin layer of a standard solution containing 20 g% bovine serum albumin. The tissue was then shock-frozen by plunging the rings into a propane/isopentane mixture (-196 °C). Cryosections of between 500 - 1000 nm thickness were cut at -90 °C using a modified Reichert cryo-ultramicrotome. The sections were sandwiched between thin collodion films and freeze-dried at -80 °C and 10^{-6} mbar in a freeze-dryer based on a turbomolecular pumping unit. Energy dispersive x-ray microanalysis of the sections was performed in a scanning electron microscope (Cambridge S150) to which a solid-state x-ray detecting system (LINK Systems) was attached. Routine measuring conditions were 20 kV acceleration voltage, between 0.2 - 0.5 nA beam current (determined in a Faraday cup at the level of the specimen), and 100 sec analysis time. Analyses were obtained in reduced raster mode, scanning areas of between .1 and .5 μm^2. The emitted x-rays were analyzed in the energy range from 0 to 10 keV, encompassing the Ka lines of the biologically relevant light elements Na, Mg, P, S, Cl, K, and Ca.

Quantification of the cellular element concentrations and cellular dry weight fractions was achieved by a comparison of the cellular x-ray spectra with those obtained in the adherent albumin layer. Concentrations are expressed as mmol/kg wet weight or, after recalculation based on the cellular dry weight fraction, as mmol/l cell water or mmol/kg dry weight. Cellular dry weight fractions are given as g% (g dry matter/100 ml). Statistical significance was evaluated by Student's t-test.

Rabbit cortical collecting duct. Isolated tubules were obtained from female New Zealand White rabbits which were kept on a standard diet and weighed between 900 and 1500 g on the day of the experiment. Rabbits were killed by decapitation and the left kidney quickly removed by abdominal incision. 1.5 - 3 mm long segments of the cortical collecting tubule were dissected in isotonic Krebs Ringer's containing (in mM) 115 NaCl, 25 $NaHCO_3$, 5 KCl, 1.2 $MgSO_4$, 1.2 Na_2HPO_4/NaH_2PO_4 (pH 7.4), 1 $CaCl_2$, 5.5 Glucose, 10 Na-acetate, and 6% bovine serum albumin. All solutions were pre-equilibrated with a mixture of 95% oxygen and 5% carbon dioxide.

Tubules were perfused using a microperfusion system which allowed freezing of the isolated tubule while being perfused. The control perfusate had an osmolarity of 290 mOsm and contained (in mM) 150 NaCl, 5 KCl, 1.2 $MgSO_4$, 1.2 Na_2HPO_4/NaH_2PO_4 (pH 7.4), and 1 $CaCl_2$. Perfusion pressure was 20 - 30 cm H_2O. The serosal bathing medium was Krebs-Ringer's (without albumin). Water flow was established by luminal perfusion with a medium of reduced NaCl concentration (130 mOsm) and serosal addition of 1 mU/ml ADH (Arginine vasopressin, Sigma). Experiments were performed at room temperature except for an initial recovery period after dissection of 60 - 90 min at 37 °C.

During the last two minutes of incubation 6% albumin was added to the perfusate. The bathing medium was rapidly aspirated and the exposed tubule snap frozen by a jet-stream of liquified propane or ethane (-180 °C). Sections of approximately 300 nm thickness were cut at -140 °C in a Reichert FC4D cryoultramicrotome and afterwards freeze-dried at -90 °C and 10^{-7} mbar. Analyses were performed in a scanning electron microscope

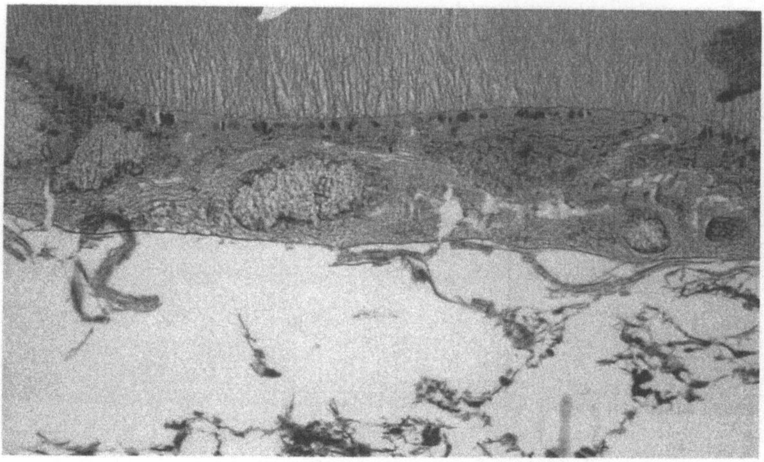

Figure 1: Scanning transmission electron micrograph of an approximately 500 nm thick freeze-dried cryosection of the toad urinary bladder epithelium (*control*). The structure on top of the epithelium corresponds to the albumin standard layer used for quantification. The height of the epithelium is approx. 6 μm.

(Cambridge S250) which was equipped with a newly designed transmission stage. Other details of the preparation are identical to those described above.

Results and Discussion

Toad urinary bladder. The investigations were performed on fresh, freeze-dried cryosections as shown in Figure 1. The epithelium consists mainly of two cell types: granular cells, which show an abundance of intracellular granules, and basal cells. Occasionally, however not present in this section, mitochondria-rich cells and goblet cells can be observed. The homogeneous layer, adjacent to the apical surface of the epithelial cells, corresponds to the layer of an albumin standard solution which was applied immediately prior to freezing. In separate experiments it was established that application of the albumin standard does not interfere with the physiological state of the epithelium.

In a first set of experiments we exposed the serosal aspect of the epithelium to half-strength Ringer's solution for about 10 min to induce cellular swelling. The incubation was done in the absence of ADH, so that the apical membrane remained essentially impermeable to water. Under this condition the cells should equilibrate only with the serosal bath. The left panel of Figure 2 shows the concentration profile for the sum of the three main intracellular ions, Na, K, and Cl. Neither in control (open symbols) nor after serosal dilution (filled symbols) did the measurements provide any evidence for an intracellular concentration gradient. In further experiments the effect of serosal dilution was tested in the presence of ADH providing nearly identical results (right panel of Figure 1). Thus, it appears that even in the presence of ADH the cells equilibrate mainly with the serosal bath.

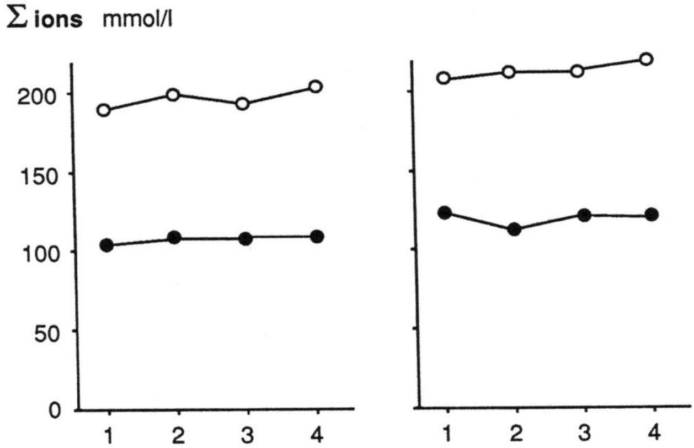

Figure 2: Intracellular profile of the sum of the intracellular ion concentrations (Na+K+Cl) in granular cells under isosmotic conditions (open symbols) and after replacing the serosal bath with half-strength Ringer's solution (filled symbols). The left panel shows an experiment without ADH, the right panel an experiment with ADH. The points 1 through 4 correspond to 4 equally spaced measuring positions along the apical-to-basal axis of the cell. From Rick and DiBona, 1987.

Table 1 summarizes the results of 59 (control) and 60 (serosal dilution) measurements obtained from three pairs of hemibladders. The concentrations are expressed in mmol/kg. w.w. and, after recalculation based on the cellular dry weight fraction, in mmol/l cell water or mmol/kg dry matter. Serosal dilution by half results in a large fall in the water concentration of all three ions; in fact, the sum of the concentration of Na, K and Cl is almost exactly halved. Under isotonic conditions the sum of Na, K and Cl is about 202 mmol/l, which is only slightly lower than the osmolarity of the Ringer's solution (220 mOsm). This finding suggests that the three ions account for most of the intracellular osmolarity and that changes in the sum of the ion concentrations can be taken as a reliable indicator of intracellular osmolarity.

The fall in the intracellular ion concentrations is partially due to cellular swelling as suggested by a reduced cellular P concentration (from 129.1 to 93.4 mmol/kg w.w.) and reduced dry weight fraction (from 19.5 to 13.9 g%). Additionally, as evident from the change in the dry weight concentrations listed in Table 1, there is a significant net loss of intracellular ions, consistent with a regulatory volume decrease of the epithelium. The cells mainly lose Cl, and to a lesser extent K and Na.

Having established that the sum of the intracellular ion concentrations can be used to estimate intracellular osmolarity changes, we performed a second series of experiments in which we induced transepithelial water flow. Figure 3 illustrates two experiments in which transepithelial water flow was stimulated by ADH and replacing the mucosal bath

Table 1: Effect of serosal dilution on the intracellular concentrations of Na, K, and Cl of granular cells.

		Na	K	Cl	Sum
Control	*mmol/kg w.w.*	12.5	116.7	33.5	162.7
	mmol/l	15.5	145.0	41.6	202.1
	mmol/kg d.w.	64.1	598.4	171.8	834.4
Serosal Dilution	*mmol/kg w.w.*	5.4	73.9	13.1	92.4
	mmol/l	6.3	85.8	15.2	107.3
	mmol/kg d.w.	38.8	531.7	94.2	664.7

Water concentrations are calculated by multiplying the wet weight concentration with 100/(100-d.w.). Dry weight concentrations are calculated by multiplying the wet weight concentration with 100/(d.w.).

with a 1/10-strength Ringer's solution . In the left experiment, the experimental hemibladder (filled symbols) was compared to a hemibladder which received only ADH (open symbols). On the right, the experimental hemibladder was compared to a hemibladder which was was exposed to the same osmotic gradient, but was not stimulated by ADH. It is evident that only in the presence of transepithelial osmotic water flow was there a detectable intracellular concentration gradient, with the most apical measurement indicating a significant intracellular dilution. On an average, using the left protocol, the intracellular dilution amounted to 29%, while using the right protocol the dilution averaged 19%.

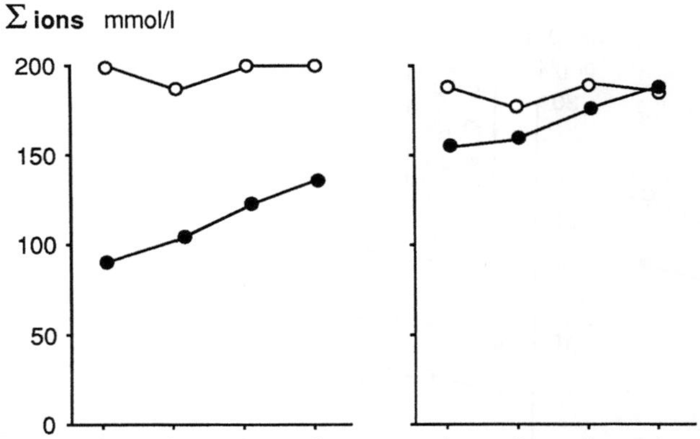

Figure 3: Intracellular profile of the sum of the intracellular ion concentrations (Na+K+Cl) in granular cells in the absence (open symbols) and presence of mucosal-to-serosal water flow (filled symbols). The points 1 through 4 correspond to 4 equally spaced measuring positions along the apical-to-basal axis of the cell. From Rick and DiBona, 1987.

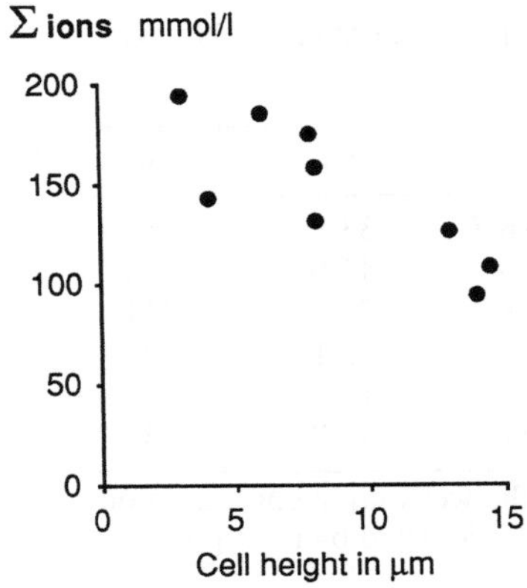

Figure 4: Correlation between the sum of the intracellular ion concentrations (Na+K+Cl) and epithelial cell height of granular cells. Data from one individual hemibladder in which osmotic water flow was induced by ADH and mucosal dilution. From Rick and DiBona, 1987.

The concentration profiles observed during osmotic water flow showed a large variability which may be accounted for, at least in part, by an effect of the epithelial cell height. This dependency is illustrated in Figure 4. Despite a considerable scatter of the data there is a clear negative correlation between the intracellular ion concentration and the epithelial

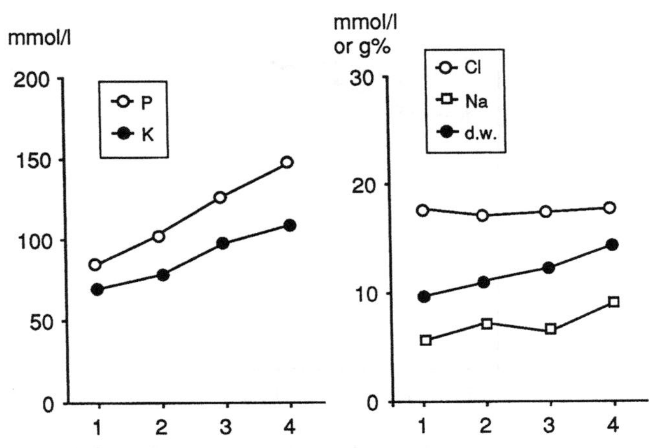

Figure 5: Intracellular concentration profiles in a granular cell during ADH-induced osmotic water flow. The points 1 through 4 correspond to 4 equally spaced measuring positions along the apical-to-basal axis of the cell. From Rick and DiBona, 1987.

cell height: the taller the cells, the larger the intracellular dilution. Three factors come to mind which may explain such a correlation. First, the cytoplasmic pathway through taller cells is longer and, therefore, should provide more resistance to water flow. Second, in the urinary bladder epithelium the cell height is mainly a function of stretch. Tall cells, therefore, should experience more water flow per nominal surface area. And third, it is conceivable that the long and tortuous lateral intercellular spaces between tall cells do not fully equilibrate with the serosal bath .

Figure 5 depicts the intracellular distribution of the three ions and of P and the dry weight in a granular cell which exhibited a large intracellular gradient (same experiment as in Figure 3, left panel). The gradient in the sum of the ion concentrations is largely attributable to a gradient of the two main intracellular cations, Na and K, whereas the anion Cl often showed no gradient, or occasionally even a reversed gradient. Consistent with the apical distension observed in light and electron microscopic sections, the P concentration and dry weight fraction in the apical cytoplasm are significantly lower than in the basal cytoplasm.

Rabbit cortical collecting duct. Investigation of this question in the mammalian cortical collecting duct is made difficult by the fact that the luminal compartment is very small and, thus, intracellular gradients can be expected to be dissipated as soon as the luminal perfusion is stopped. We, therefore, developed a microperfusion system which allowed freezing of the tubule while being perfused and rapid removal of the tissue to prevent re-warming after freezing. The tubule was visualized using differential interference contrast microscopy which made it possible to judge the viability of the tissue and to monitor

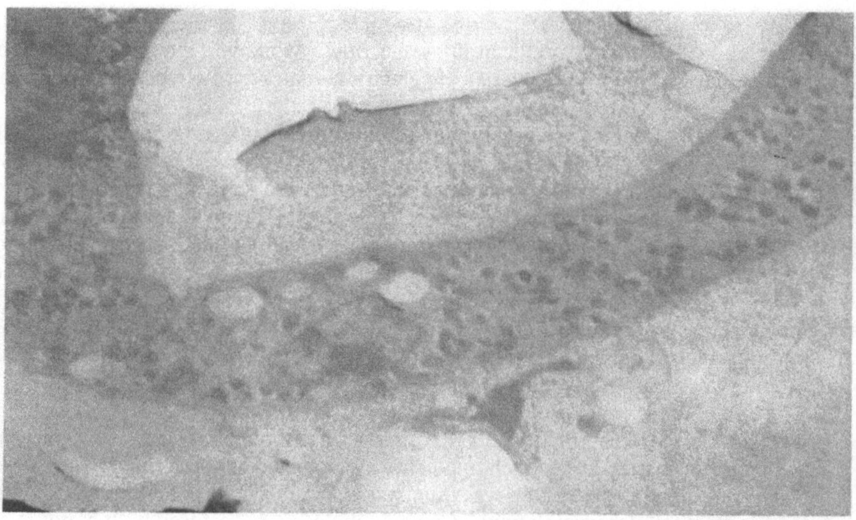

Figure 6: Scanning transmission electron micrograph of a 300 nm thick freeze-dried cryosection of the rabbit cortical collecting duct (*control*). The cell in the upper left corner probably represents an intercalated cell. The field of view is approx. 20 μm wide. For further details see text.

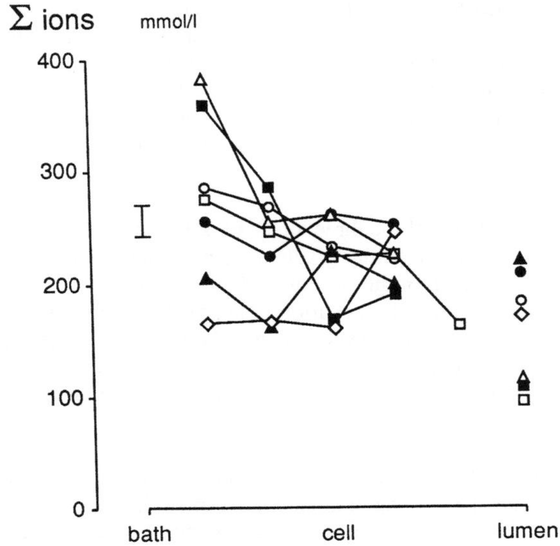

Figure 7: Intracellular profile of the sum of the intracellular ion concentrations (Na+K+Cl) in principal cells during ADH-induced osmotic water flow. In addition, the results of measurements in the tubular lumen and bath are depicted. Different symbols denote different sections. Data combined from 3 different tubules.

swelling and vacuolation of cells during water flow. Moreover, the transepithelial potential difference was recorded to verify the electrical (natriferic) effect of the hormone.

Figure 6 shows a freeze-dried cryosection of the cortical collecting duct. In most sections, as in this one, the tubule appears compressed, probably due to the impact of the jet-stream of coolant. The light ground substance of bath and lumen corresponds to albumin which was added immediately prior to freezing in order to prevent dislocation of ions during freeze-drying. In most sections two different types of cells could be distinguished: clear cells which, in particular during osmotic water flow, showed apical vacuolations and dark cells which were characterized by a more electron-dense cytoplasm. According to frequency and size and distribution of intracellular organelles, clear cells most likely represent principal cells, and dark cells represent intercalated cells as described in conventional preparations (Madsen et al., 1988; Ridderstrale et al., 1988).

The tissue was frozen approximately 40 min after transepithelial water flow was established by perfusing the lumen with hypotonic medium and addition of ADH to the serosal bath. At this time the epithelium was usually maximally swollen and the cells showed extensive intracellular vacuolations. Control tubules were perfused with hypotonic medium only, exposed to ADH only, or exposed to neither experimental perturbation.

In none of the controls, but in most of the experimental tubules, an intracellular ion concentration gradient was detectable. Figure 7 combines the results from several tubules which were exposed to transepithelial water flow. In most profiles, a basal-to-luminal gradient of the sum of Na, K, and Cl concentrations is apparent. The figure also shows the sum of the ion concentrations in the tubular lumen. In about half of the sections the observed luminal osmolarity was significantly higher than the nominal osmolarity of the perfusate indicating that, at the level of the section, a significant fraction of the water was already absorbed. The variability of the luminal composition may, at least in part, explain the large variability of the cellular gradients.

Summary. In both epithelia, a significant intracellular ion concentration gradient was observed when transepithelial flow of water was established. To the extent that intracellular ions are osmotically active, the data suggest that a sufficiently brisk cytoplasmic flow of water can induce intracellular osmolarity gradients, or gradients in water activity. The presence of these gradients implies that, in addition to the two limiting cell membranes, the cytoplasmic pathway itself has to be considered as a barrier to the transepithelial flow of water.

The apical-basal distribution of P and dry matter may be explained by the fact that large mobile macromolecules are dragged along with the water flow. Moreover, friction between cytoskeletal elements and water could result in a slight pressure build-up in the apical pole of the cell and distension of the cytoskeletal matrix. The concentration gradients for small ions cannot be accounted for by solvent drag alone. More likely, they are the result of an uneven distribution of fixed negative charges as induced by the water flow. Such a Donnan-type distribution would also explain why the distribution of Cl showed a different concentration profile.

References

DiBona DR, Civan MM, Leaf A (1969) The cellular specificity of the effect of vasopressin on toad urinary bladder. J Membrane Biol 1:79-91

DiBona DR (1981) Vasopressin action on the conformational state of the granular cell in the amphibian urinary bladder. In: Epithelial Ion and Water Transport. ADC Macknight and JP Leader, editors pp. 241-255 Raven, New York

Ganote CE, Grantham JJ, Moses HL, Burg MB, Orloff J (1968) Ultrastructural studies of vasopressin effect on isolated perfused renal collecting tubules of the rabbit. J Cell Biol 36:355-367

Kirk KL, Schafer JA, DiBona DR (1984) Quantitative analysis of the structural events associated with antidiuretic hormone-induced volume reabsorption in the rabbit cortical collecting tubule. J Membrane Biol 79:65-74

Madsen KM, Verlander JW, Tisher CC (1988) Relationship between structure and function in distal tubule and collecting duct. J Electr Microsc Tech 9:187-208

Peachey LD, Rasmussen H (1961) Structure of the toad urinary bladder as related to its physiology. J Biophys Biochem Cytol 10:529-553

Rick R, Dörge A, Thurau K (1982) Quantitative analysis of electrolytes in frozen dried sections. J Microsc (Oxford) 125:239-247

Rick R, DiBona DR (1987) Intracellular solute gradients during osmotic water flow: An electron-microprobe analysis. J Membrane Biol 96:85-94

Ridderstrale Y, Kashgarian M, Koeppen B, Giebisch G, Stetson D, Ardito T, Stanton B (1988) Morphological heterogeneity of the rabbit collecting duct. Kidney Int 34:655-670

Schratt W, Rick R, Walker JT, DiBona DR (1988) Intracellular osmolarity gradients in isolated perfused cortical collecting tubules during ADH-induced water flow. Proc Am Soc Nephrol (in press)

Conference Discussion

G.M. Roomans: Your data on elemental gradients across the bladder epithelium show that the gradient for P, but not that for Cl, parallels the gradients for Na and K. Would that imply that the distribution of (fixed) phosphate groups determines the cation distribution within the cell?

R. Rick: Yes. At physiological pH, a significant fraction of the intracellular posphate groups is dissociated. Given the large intracellular phosphate concentration, one would expect that phosphate groups represent the largest fraction of fixed negative charges.

Th. v. Zglinicki: Your result on toad urinary bladder demonstrates nicely some kind of binding or restricted mobility of ions due to the interaction with cellular polyelectrolytes. Should such a restriction of ion mobility not have consequences for, e.g., osmotic activity?

R. Rick: The higher concentration of polyelectrolytes in the basal portion of the cytoplasm may, indeed, lower the chemical and osmotic activity of small cations, such as Na and K. It is unlikely, however, that this effect would completely offset the osmotic imbalance created by an uneven distribution of fixed charges. In Donnan systems, water generally moves from the site of lower concentration of fixed polyelectrolytes to that of higher concentration.

B.L. Gupta: I am pleased to see that you have found gradients of Na, K, and Cl in these granular cells of toad bladder. We reported similar gradients in sugar-stimulated rabbit ileum where enterocytes were absorbing fluid from the apical side. In these 40/um tall, structurally much more complex cells, the gradients in the main cell body were different from those in the organelle-free peripheral zone (Fed. Proc. 1979; Benzon Symp. 1981).

How far is it justified to add Na+K+Cl for estimating intracellular osmolality? It would be much more appropriate to consider (Na+K) times 2, since anions and cations in free solution (osmotically active) must also be balanced. Anions may be free organic solutes.

R. Rick: We have chosen the sum of the three ions as this involved the least number of assumptions. Given that many negative charges are polyanionic, it is unlikely that the osmotic concentration of negative charges equals that of positive charges. In my view, it is not possible to cleanly distinguish between truly 'free' and 'bound' intracellular solution, at least not in our measurements.

I might add that our data are not entirely comparable to your data as the rate of isotonic fluid transport in the small intestine is about one order of magnitude less than the rate of osmotically induced water flow in ADH-stimulated epithelia.

Characterization of Electrolyte Transport Mechanisms and Compartments by the Use of the Markers Rb and Br*

A. Dörge, F.X. Beck, R. Rick**, W. Nagel, K. Thurau

Physiologisches Institut
der Universität München
Pettenkoferstr. 12
8000 München 2
FRG

INTRODUCTION

The application of electron microprobe analysis to biological tissues allows the determination of electrolytes within individual cells or even subcellular structures. This capability enables the elucidation of the transport pathways for different electrolytes in different cell types of an epithelium. One way to clarify transepithelial cellular transport routes of ions is the exchange of the ions in question against marker ions which are not common in biological fluids but which are similarly transported. If the marker ion is added to that side of the epithelium from which the flux originates those cells involved in the transport should be marked. Furthermore marker ions can also be used to elucidate complex transport processes, such as cotransports, across a single cell membrane. During recent years we have used the markers Br and Rb to characterize transport mechanisms in several epithelial tissues. Rb instead of K was employed to elucidate the transport characteristics of a basolaterally located electrolyte cotransport system involved

* Supported by grants from the Deutsche Forschungsgemeinschaft

** University of Alabama in Birmingham
Department of Physiology and Biophysics
Nephrology Research and Training Center
University Station
Birmingham, AL 35294
USA

in cellular volume regulation of amphibian skins. Br was substituted for Cl in the apical and basal bathing solutions of toad skin to gain further information on the route of transcellular Cl transport. Using Rb in the luminal bathing solution of rabbit urinary bladder, it was possible to localize the Na transport compartment in this multilayered epithelium.

METHODS

Abdominal skins of Rana esculenta and Bufo viridis and rabbit urinary bladders were mounted in Ussing-type chambers. After the short-circuit current had reached a steady state value, during which time both sides of the epithelia were bathed with normal Ringer's solution, the incubation conditions were changed according to the protocols described in Results and Discussion.

At the end of the incubation the epithelium was quickly removed from the chamber and prepared for electron microprobe analysis. After covering the apical side with a thin layer of an albumin standard solution (20 g bovine albumin dissolved in 80 ml apical bathing solution), the epithelia were frozen in a propane/isopentane mixture at liquid N_2 temperature. Sections of about 1 um thickness were cut from the frozen tissue in a modified Reichert cryoultramicrotome (Reichert OmU 3, Vienna, Austria) at $-90^{o}C$, sandwiched between thin films of collodion and formvar and freeze-dried at $-80^{o}C$ and 10^{-6} mbar.

X-ray microprobe analysis was performed in a scanning electron microscope (Stereoscan S 150, Cambridge Instruments, UK) with an energy dispersive X-ray detector (LINK Systems, High Wycombe, UK). For analysis, areas of about 1 um^2 were scanned for 100 sec within the cells and the albumin standard layer. The acceleration voltage used was 20 kV and the probe current 0.3 nA. Discrimination between the characteristic elemental X-ray peaks and the background was performed with a special computer program (Bauer and Rick 1978). Quantification of the cellular elemental concentrations in mmol/kg wet weight (w.w.) was achieved for each section by comparing the intensities of the characteristic elemental radiations obtained in the cells with those of the albumin standard layer. The cellular dry weight content was determined by comparing the background radiations of cell and standard spectra. Further details of tissue preparation, analysis and quantification procedure have been given previously (Dörge et al. 1978).

RESULTS AND DISCUSSION

Na-K-2Cl cotransport in principal cells of frog skin

Figure 1 shows some essential transport features of the principal cells in the frog skin epithelium. Beside the constituents involved in transepithelial Na transport (channels for passive Na entrance across the apical membrane and Na/K pumps for active Na extrusion across the basal membrane) the basal membrane is equipped with an electroneutral Na-K-2Cl cotransport system which maintains the cellular Cl concentration far above its electrochemical equilibrium (Dörge et al. 1985; Ussing 1985). The existence of such a cotransporter was deduced from measurements of the cellular volume and Cl concentration. For example, incubating the basal side with Na- or K-free Ringer's solutions

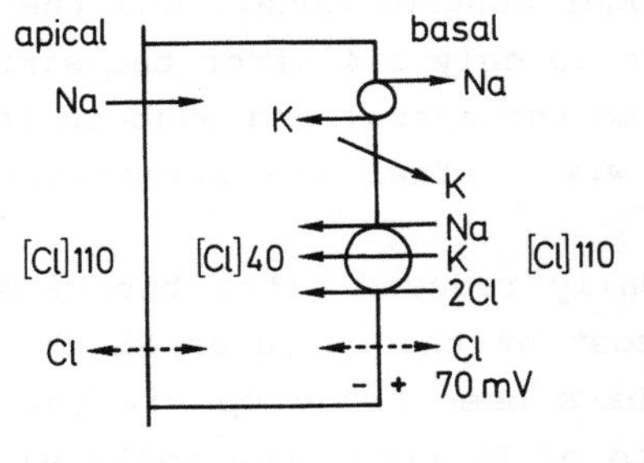

Figure 1: Transport features of apical and basal membranes of principal cells in frog skin. Also the cellular Cl concentrations of the different compartments and the electrical potential of the basal membrane are given. For details see text.

results in a decrease of cellular Cl concentration and cell shrinkage (Ussing 1965; Ferreira and Ferreira 1981). Recovery of cell volume and Cl concentration after Cl depletion by incubating the basal side with either Cl-free or half-isotonic Ringer's solution is critically dependent on the presence of Na and K in the Ringer's solution on the basal side and can be prevented by cotransport inhibitors (Ussing 1985; Dörge et al. 1985). Since the cells loose Cl and shrink after they have been expanded by application of half-isotonic Ringer's solution to the basal side, Ussing concluded that the Cl transport system is involved in cellular volume regulation (Ussing 1982).

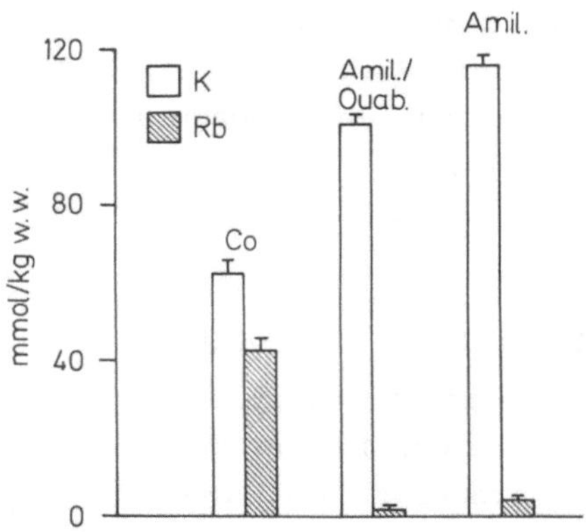

Figure 2: K and Rb concentrations of principal frog skin cells after 1 h of basal incubation with Rb-Ringer's solution under control conditions (Co) and after blocking the transepithelial Na transport either with amiloride and ouabain or amiloride. Drug concentrations were 10^{-4} M. (Mean values of 3 experiments ± 2 SE).

In order to further study the characteristics of the cotransport system the cellular uptake of Rb was measured under various experimental conditions. Figure 2 shows the cellular K and Rb concentrations under control conditions and after blocking the transepithelial Na transport. In each case the Rb-Ringer's solution, in which all K (2.5 mM) was replaced by Rb, was applied for 1 h to the basal side. Whereas about 40 % of the cellular K is replaced by Rb under control conditions the cellular Rb concentrations increase to only 1.4 after the simultaneous application of amiloride to the apical and ouabain to the basal side and to 4.2 mmol/kg w.w. after the application of amiloride alone.

Since the Rb uptake is drastically reduced after blocking the transepithelial Na transport, most of the cellular Rb observed under control conditions must have been taken up via the Na/K pump and thus only minor amounts of Rb enter the cells via the basal K channels or the Na-K-2Cl cotransporter. This can be interpreted such that the cotransporter is almost completely silent under control conditions and that the Cl permeability of both apical and basal membrane must be very low to keep the cellular Cl concentration above electrochemical equilibrium. Similar results regarding the Cl permeabilities have also been obtained by cellular volume measurements (Ussing 1985). The somewhat larger increase in the cellular Rb concentration after amiloride compared with the simultaneous application of amiloride and ouabain might reflect a residual activity of the

Na/K pump maintained by a small Na influx across the basal cell membrane.

Figure 3 summarizes the major results obtained after basal incubation with Cl-free gluconate- and subsequently with Rb-Ringer's solutions. Ouabain was applied throughout the experiments to the basal side to avoid cellular Rb uptake by the Na/K pump and amiloride to the apical side to prevent drastic changes in the cellular electrolyte concentrations due to the inhibition of the Na/K pump. After Cl-free incubation the cellular Cl concentration is lowered to 20.7 mmol/kg w.w. and subsequently increases during 1 h of Rb-Ringer's solution to 35.6 mmol/kg w.w. Cellular Rb concentration increases simultaneously to 30.4 mmol/kg w.w. From the changes in the cellular Cl and Rb concentrations and the dry weight contents an increase in the cellular Cl and Rb of 80.3 and 127.6 mmol/kg dry weight can be calculated.

The finding that the gain in cellular Rb is larger than that of Cl during the Cl reuptake phase is difficult to interpret. Since it was shown that no substantial Rb amounts enter the

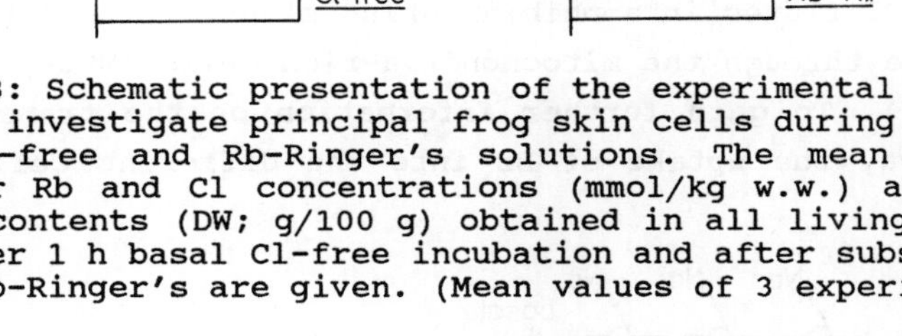

Figure 3: Schematic presentation of the experimental conditions used to investigate principal frog skin cells during incubation with Cl-free and Rb-Ringer's solutions. The mean values of cellular Rb and Cl concentrations (mmol/kg w.w.) and the dry weight contents (DW; g/100 g) obtained in all living cell layers after 1 h basal Cl-free incubation and after subsequent 1 h basal Rb-Ringer's are given. (Mean values of 3 experiments).

cells via the K channels, we concluded that during the Cl reuptake phase both Cl and Rb are taken up by the cotransport system. To explain the pronounced cellular Rb accumulation it was considered that the cotransporter might work like a shuttle. Assuming this to be so the Cl reuptake should be retarded, whereas Rb is transported into and K out of the cells so that only the Rb uptake reflects the inwardly directed transport rate of the cotransporter. However, recent Cl reuptake measurements made in the presence of bumetanide raise some doubt as to this interpretation. Although the cotransport inhibitor blocked the Cl reuptake almost completely, the cellular uptake of Rb was not influenced. This result could be explained if it were assumed that the subunit of the cotransporter to which Rb and K bind works independently of the other subunits. Then the question arises, however, why such an exchange of cellular K against Rb is not observed under control conditions. Further investigations are necessary to answer the question as to whether the Rb accumulation during the Cl reuptake phase does, in fact, take place via the cotransporter and if so, how this can be understood on a molecular basis.

Transepithelial Cl pathways in amphibian skins

Figure 4 shows schematically the pathways involved in transepithelial NaCl reabsorption of amphibian skins. Whereas it is generally accepted that the Na transport is accomplished by the principal cells, which all together form a syncytial Na transport compartment (Ussing and Windhager 1964; Rick et al. 1978), it seems also clear that the electrodiffusional Cl movement, which under in vivo conditions must be involved in the NaCl reabsorption, follows a different route. During the last 10 years several studies have provided indirect evidence that passive Cl fluxes in amphibian skins might follow a transcellular route through the mitochondria-rich cells (MRCs) (Larsen et al. 1987). To gain further information on the transepithelial Cl pathway the uptake of Br into the different cell types of

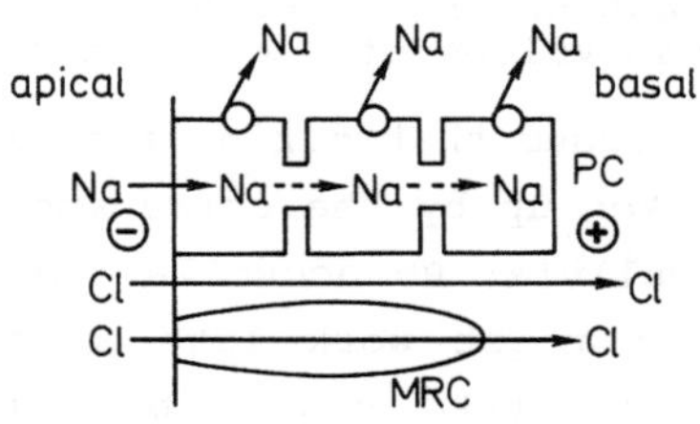

Figure 4: Schematic presentation of transepithelial routes of Na and Cl in amphibian skins. PC = principal cells; MRC = mitochondria-rich cell.

toad skin was determined under control conditions and after activating the transepithelial Cl flux by basal positive clamp potentials. Br was chosen as a marker since Harck and Larsen (1986) demonstrated that Br is transported similarly to Cl. Figure 5 shows the transepithelial current under short-circuited conditions and after perturbation of the transepithelial potential from 0 to 100 mV. Whereas the positive current observed under short-circuited conditions (+ 15 uA/cm^2) is mainly due to active inward Na transport, the large negative current (-250 uA/cm^2) developed after voltage perturbation is mainly carried by an inwardly directed passive Cl flux (Larsen 1982).

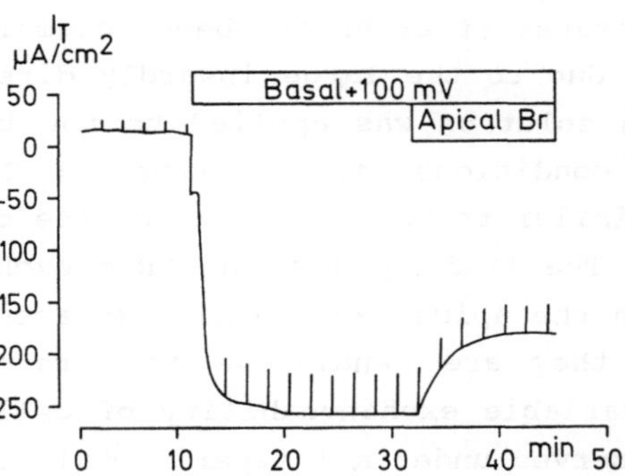

Figure 5: Time course of the transepithelial current of toad skin after perturbing transepithelial potential from 0 to 100 mV (basal side positive) and after changing the apical incubation medium from normal to NaBr-Ringer's. The heights of the current deflections reflect the transepithelial conductance.

Substituting Br for Cl in the apical bathing solution decreases the negative current by only 30 %. Figure 6 demonstrates the

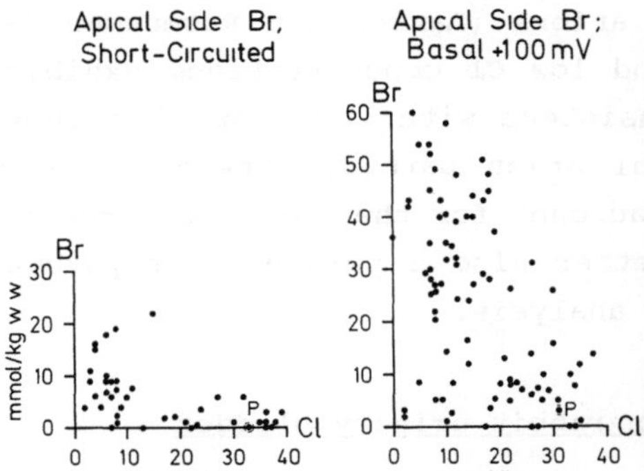

Figure 6: Cellular Br against Cl concentrations after incubating the apical side with Br-Ringer's for 15-30 min under short-circuited and for 5-30 min under voltage-activated conditions (basal side +100 mV). Given are the values of individual MRCs and the mean values of principal cells marked by the symbol "P". (Values of 7 experiments).

cellular Cl/Br exchange of individual MRCs after applying the Br-Ringer's solution under short-circuited and voltage activated conditions to the apical side. The mean values obtained in the principal cells are also included (symbol "P"). Compared to the short-circuited state a certain portion of the MRCs contains far higher Br concentrations after voltage activation. However, it is also obvious that under both experimental conditions many MRCs do not exchange Cl with Br to any substantial extent. No Br is detectable in the principal cells under short-circuited conditions and only small amounts of Br can be found in this cell population after voltage activation. Thus these cells seem not to be involved in transepithelial anion transport. The small amounts of Br taken up into this cell type during voltage activation could be explained by a Cl/Br exchange across the basal membranes after Br has been accumulated in the intercellular spaces due to the large inwardly directed Br flux. If the Br-Ringer's solution was applied to the basal side under short-circuited conditions, the exchange of Cl in the MRCs with Br was very similar to that observed on the opposite side (data not shown). The finding that the MRCs exchange their Cl with Br across both the apical and basal membranes can be taken as evidence that they are candidates for transepithelial Cl transport. The variable exchangeability of cellular Cl with extracellular Br observed under all experimental conditions, however, suggests that the MRCs are a heterogeneous population in regard to anion transport. Under voltage activated conditions, when the apical membrane of MRCs is assumed to be highly permeable to anions (Larsen and Rasmussen 1985), only cells with high Br and low Cl concentrations exhibit exchange characteristics consistent with the view that they represent the transepithelial anion pathway. Whether the small number of these cells can account for the observed large transepithelial current, or whether also a paracellular pathway is involved, requires further analysis.

<u>Na transport compartment in rabbit urinary bladder</u>

During recent years evidence has been accumulated indicating that the rabbit urinary bladder possesses transport features for active Na reabsorption similar to those already described for other tight epithelia (Lewis and Hanrahan 1985). Based on cellular potential and membrane capacitance measurements (Lewis et al. 1976; Clausen et al. 1979) it was concluded that only the apical cell layer is involved in transepithelial Na transport as illustrated in figure 7 A. Since this view seems to conflict with our previous findings that multilayered epithelia like the frog skin (Rick et al. 1978) behave like a functional syncytium in regard to transepithelial Na transport, experiments were conducted to investigate this problem. Initially,

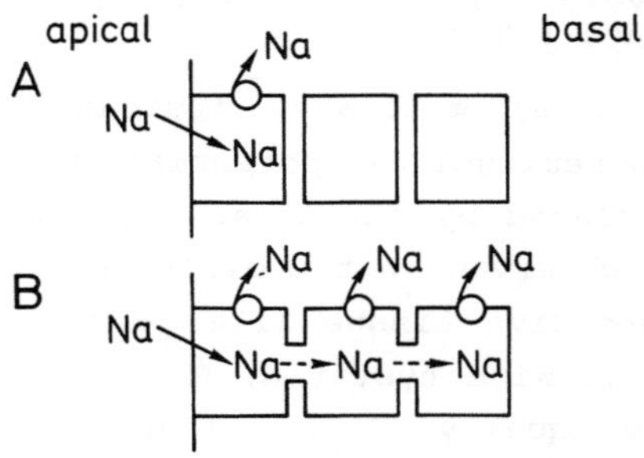

Figure 7: Schematic presentation of the Na transport compartment in rabbit urinary bladder. A: only the apical cell layer is involved. B: all cell layers form a syncytial Na compartment.

experiments similar to those in the frog skin were performed to localize the Na transport compartment. In frog skin strong evidence for a syncytial Na transport compartment was obtained by the fact that the increase in cellular Na concentration observed after blocking the Na exit step with ouabain was prevented in all epithelial cell layers by incubating the apical side with Na-free Ringer's solution. However, such an inhibition of the ouabain induced Na increase could not be observed in the rabbit urinary bladder. It therefore appears that in contrast to the frog skin the basal Na permeability is much higher than the apical one. Thus, such experiments thus yielded no clear-cut answer as to the localization of the Na transport compartment in this multilayered epithelium.

In the second attempt to identify the Na transporting cells, Rb, was used as a marker on the apical side. Before applying Rb, the apical membrane of the luminal cell layer was permeablised for anions and cations by nystatin (Lewis et al. 1977). To avoid uncontrolled changes in the cellular electrolyte composition due to the opening of the apical membrane, nystatin was given to the apical side in a "cellular solution" which mainly contained K_2SO_4 (25 mM), $KHCO_3$ (25 mM), and sucrose (160 mM), but no Na, Cl, and Ca. After the insertion of the

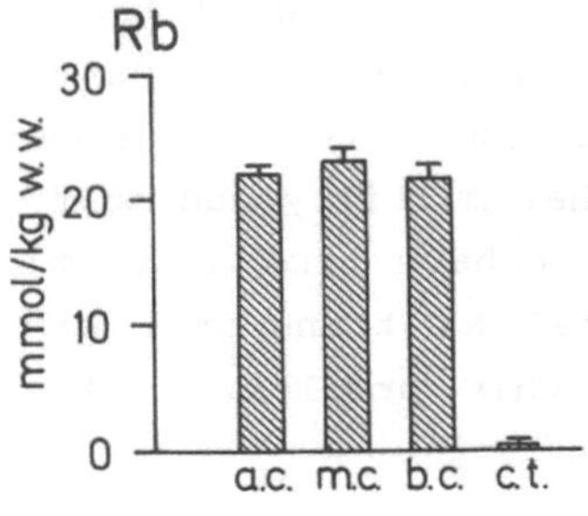

Figure 8: Uptake of Rb into the different epithelial cell layers and connective tissue of rabbit urinary bladder after incubating the apical side for 6 min with a 50 mM Rb and 30 mg/l nystatin containing solution. a.c., m.c., b.c. and c.t. = apical, middle, and basal cell layer, and connective tissue, respectively. (Mean values of 2 experiments ± 2 SE)

nystatin channels into the apical membrane, indicated by a drastic increase in the transepithelial potential to about 50 mV, 25 mM of K_2SO_4 were replaced by equimolar quantities of Rb_2SO_4. The Rb concentrations observed in the different epithelial cell layers and the connective tissue after 6 min of Rb application are given in Fig 8. With nystatin, Rb is taken up by all epithelial cell layers equally. Thus, it must be concluded that the rabbit urinary bladder epithelium can also be regarded as a syncytial Na transport compartment, similar to that described for the frog skin and shown in Fig. 7 B.

REFERENCES

Bauer R, Rick R (1978) Computer analysis of X-ray spectra (EDS) from thin biological specimens. X-ray Spectrom 7:63-69

Clausen C, Lewis SA, Diamond JM (1979) Impedance analysis of a tight epithelium using a distributed resistance model. Biophys J 26:291-318

Dörge A, Rick R, Gehring K, Thurau K (1978) Preparation of freeze-dried cryosections for quantitative X-ray microanalysis of electrolytes in biological soft tissues. Pflügers Arch. 373:85-97

Dörge A, Rick R, Beck F, Thurau K (1985) Cl transport across the basolateral membrane in frog skin epithelium. Pflügers Arch. 405:S8-S11

Ferreira KTG, Ferreira HG (1981) The regulation of volume and ion composition in frog skin. Biochim Biophys Acta 646:193-202

Harck AF, Larsen EH (1986) Concentration dependence of halide fluxes and selectivity of the anion pathway in toad skin. Acta Physiol Scand 128:289-304

Larsen EH (1982) Chloride current rectification in toad skin. In: Zadunaisky J (ed) Chloride transport in biological membranes. Academic Press, Inc, New York, p 333-364

Larsen EH, Rasmussen BE (1985) A mathematical model of amphibian skin epithelium with two types of transporting cellular units. Pflügers Arch 405:S50-S58

Larsen EH, Ussing HH, Spring KR (1987) Ion transport by mitochondria-rich cells in toad skin. J Membrane Biol 1:25-40

Lewis SA, Eaton DC, Diamond JM (1976) The mechanism of Na^+ transport by rabbit urinary bladder. J Membrane Biol 28:41-70

Lewis SA, Eaton DC, Clausen C, Diamond JM (1977) Nystatin as a probe for investigating the electrical properties of a tight epithelium. J Gen Physiol 70:427-440

Lewis SA, Hanrahan JW (1985) Apical and basolateral membrane ionic channels in rabbit urinary bladder epithelium. Pflügers Arch. 405:S83-S88

Rick R, Dörge A, v. Arnim E, Thurau K (1978) Electron microprobe analysis of frog skin epithelium: Evidence for a syncytial sodium transport compartment. J Membrane Biol 39:313-331

Ussing HH (1965) Relationship between osmotic reactions and active sodium transport in the frog skin epithelium. Acta Physiol Scand 63:141-155

Ussing HH (1982) Volume regulation of frog skin epithelium. Acta Physiol Scand 114:363-369

Ussing HH (1985) Volume regulation and basolateral co-transport of sodium, potassium, and chloride ions in frog skin epithelium. Pflügers Arch 405:S2-S7

Ussing HH, Windhager E (1964) Nature of shunt path and active sodium transport path through frog skin epithelium. Acta Physiol Scand 61:484

Conference Discussion

Th. von Zglinicki: Could recycling of Cl play a role for the high uptake of Rb in the first set of experiments? There should be a certain Cl permeability, because half of the Cl is lost in Cl-free solution.

A. Dörge: Although Cl recirculation during the Cl reuptake phase cannot completely be excluded, its existence seems implausible for the following reason: As demonstrated by the low cellular Rb uptake observed after blocking the Na transport without previous Cl-free incubation, the Na-K-2Cl cotransporter seems to be practically silent under control conditions and the Cl permeability of both apical and basal membranes must be very low to keep the cellular Cl concentration above electrochemical equilibrium. Furthermore, the cellular Cl loss during Cl-free incubation can be prevented by the cotransport inhibitor bumetanide. It, therefore, appears that under the conditions applied apart from the cotransporter no other essential transport path for net Cl movement exists across the basal cell membrane.

Electron Probe Analysis of Transport Properties of Cultured Cells

C. Lechene
Department of Medicine
Laboratory of Cellular Physiology
National Electron Probe Resource for the Analysis of Cells
Harvard Medical School and Brigham and Women's Hospital
221 Longwood Av, Boston
MA 02221, USA

Potassium, sodium and chloride cellular leak pathways and their components, (Na,K)-pump steady state rate of pumping, apparent maximum velocity, affinity for Na and affinity for other ligands (ouabain) are measured in whole cells using electron probe analysis. The general method for electron probe analysis of cultured cells has been described in several application and review articles (Abraham et al., 1985; Larsson, L. et al., 1986 ; Larsson, S. et al., 1986 ; Lechene, 1986 ; Lechene, 1988).

In this chapter we will discuss some methodological steps, then we will briefly mention a series of results obtained using electron probe analysis of cultured cells.

I. METHODOLOGY

A. TECHNIQUE

After isolation, cells are plated on silicon chips and cultured for two to four days. Small number of cells and a few days culture provide the following advantages: many (if not any) cell type may be studied; cells appear to maintain their terminally differentiated functions; there is no fibroblasts overgrowth. We have performed electron probe analysis of short term primary cultures of rat renal proximal tubules, renal medullary thick ascending limb, inner medullary collecting duct, epithelial and mesangial glomerular cells, rat hepatocytes, guinea pig gastric cells, bovine retina endothelial cells, and shark rectal gland cells.

In order to perform experiments with the same cells using both fluorescence microscopy and electron probe analysis, a light-transparent and electrical conductive backing for the cells is required. Quartz was tested for this purpose. Fibroblasts(NIH 3T3) were seeded on quartz pieces coated with thin carbon or aluminum layers, and intracellular content of P, Cl, K, and Na were measured. Results are presented in Table 1. The results are equivalent with values from control groups seeded on silicon pieces within their statistical errors (see table), which is supported by several group correlation tests.

The errors are a measure of the mean spread in composition of individual cells. The results for any type of support do not deviate significantly from those for silicon chips coated with carbon, that are ordinarily used. This holds even for Na/P ratios of aluminum-covered pieces, in spite of the high x-ray absorption coefficient of aluminum for low energy x rays. Thus, quartz coated with either carbon or aluminum can be used as a cell support (Both and Lechene, unpublished).

Table 1. Comparison of Quartz to Silicon Supports for Electron Probe Analysis of Cultured Cells.

			mole/mole	
Type of Support	N	Cl/P	K/P	Na/P
Silicon Carbon coated	11	.23 ± .06	1.12 ± .13	.11 ± .03
Silicon Al coated	8	.28 ± .03	1.15 ± .05	.08 ± .01
Quartz Carbon coated	11	.23 ± .03	1.16 ± .09	.12 ± .02
Quartz Al coated	13	.24 ± .08	1.08 ± .14	.13 ± .09

After experimental treatment cells are washed in cold distilled water or ammonium acetate for six seconds, quenched and freeze dried. In any new cell type we compare washing with distilled water to washing with isotonic ammonium acetate or isotonic NaCl (then, only K can be measured). The results of such a comparison is shown in Table 2 for green monkey kidney CV-1 cells and for OR-6 cells (CV-1 cells transfected with a 'ouabain resistance gene'). As we observed it with human diploid fibroblasts(Abraham et al., 1985) and with renal proximal tubular cells (Lechene, 1986) distilled water preserve at best intracellular content of K and Na and thus is an excellent washing medium for preparation of these cells for electron probe analysis (Epstein and Lechene, 1988, and unpublished data).

Table 2. Comparison of the Effect of Washing Media on Ionic Content in CV1 and OR6 Cells[a]

	K/Na		K/P	
	H_2O	CH_3COONH_4	H_2O	NaCl
CV1	6.09 ± .45 (34)	6.09 ± .30[b](36)	.61 ± .02 (34)	.57 ± .03[b](21)
OR6	4.7(37)	4.1[b](30)		

[a]Values are mean ± SE. for CV1 cells and mean, compared using Mann-Whitney rank sum test, for OR6 cells ; (): number of cells. [b]Not statistically different from H_2O.

Data acquisition and data manipulation are performed under computer control. Mass of cells analysed is normalised with respect of the amount of P (cellular phosphorus is in excess of cellular phosphate by about 100 fold) or of C contained in the cellular volume analysed under the electron beam.

With any new cell type it is verified that there is a strong linear relation between P and C cellular amount. In general the mean amount of phosphorus under the electron beam does not change within the extremes of a specific experiment. For example, as showed in Table 3, the mean values of phosphorus content in CV-1 and OR-6 cells (induced or not to express the transfected ouabain resistance gene) are equivalent in control cells maintaining a low intracellular Na and in cells that had loaded Na and lost their K after inhibition of the (Na,K)-pump (Epstein and Lechene, 1988, and unpublished observations).

Table 3. Cellular Phosphorus content in Control Cells (high intracellular K and after (Na,K)-Pump inhibition (high intracellular Na)[a]

	CV1	OR6 uninduced	OR6 induced
Control	93.9 ± 6.7 (22)	93.1 ± 9.1 (16)	91.5 ± 5.8 (22)
After Na loading	103 ± 4.6 (69)	91.8 ± 3.8 (45)	93.9 ± 3.9 (73)

[a]Values are mean P in femtomoles/100μm^{2}dried cell ± SE ; (): number of cells.

Spectrometers are calibrated using liquid droplets of known composition (Abraham et al., 1985). Making liquid droplet standards used for chemical elements x-ray calibration is time-consuming, and may introduce human errors. In order to decrease these factors, we used custom-designed bulk x-ray reference standards for electron probe calibration (Tousimis Research Corp., Rockville, MD). They contain the elements C, N, O, F, Na, Mg, Al, Si, P, S, Cl, K, Ca, Fe, Co, Zn, Br, Rb, Ag, I, and Cs in the most stable or most suitable form that is available (BN, C, LiF, NaCl, MgO, Al, Si, KBr, Ca5(PO4), Fe, Co, ZnS, Rubidium acid phthalate, Ag, and CsI). Test runs have demonstrated that characteristic x-ray intensity ratios for different elements are independent of beam current (20-200 nA) and beam diameter (15-75 μm) within 2%.

In order to normalize among the various spectrometers the relation between x-ray intensity and amount of element analyzed, the bulk standards' yields have been compared to those of droplet standards with known concentrations. The yield ratios of the individual elements to phosphorus were used for calculation (Table 4) (Both and Lechene, unpublished).

Except for sulphur, the standard deviations are only slightly larger than the statistical errors, which were derived from the pure counting statistics. These statistical errors do not include long term beam current drifts, which may affect the ratio element/phosphorus for the bulk standards, because in these tests, phosphorus and the element in question usually were not measured at the same time. Therefore, it is expected that the standard deviations are slightly larger than the statistical errors. These results demonstrate that the bulk standards can replace the droplet ones.

Table 4. Ratio of Droplet Standards over Bulk Standards

Droplet/Bulk	Cl/P	K/P	Na/P	Ca/P	Mg/P	S/P
Mean	12.6	.676	14.9	.695	.249	1.1
Standard Deviation (%)[a]	3.2	2.7	3.5	4.7	4.3	2.1
Statistical Error (%)	2.4	2.8	2.4	2.8	3.5	4.0

[a]Results of 5 experiments.

B. LEAK PATHWAYS AND (Na,K)-PUMP ACTIVITY

Initial rates of leaks or of (Na-K)-pump activity are derived from linear or non-linear fit of changes in ionic content with time(Harris et al., 1986 ; Lechene, 1986).

In order to verify that in normal cells (with low intracellular Na) ouabain inhibition of the pump is immediate compared to the rates of Na and K leaks, we use cells Rb loaded by preincubation in 5 mM RbCl followed by return in 5mM KCl. In this protocol initial rate of K influx is an index of pump activity and initial rate of Rb efflux is an index of leak pathway (Harris et al., 1986 ; Lechene, 1986).

As was measured in primary culture of renal proximal tubular cells (Harris et al.,1986) we verify that: 1) K or Rb influx is immediately blocked by ouabain 10^{-3} M, indicating the adequacy of ouabain inhibition of the pump when measuring initial rates of leak pathways; and 2) initial rate of K efflux measured either after ouabain inhibition of the pump or after abruptly switching the cells to RbCl containing medium are equivalent indicating that in our experimental conditions, the initial rate of K efflux is not measurably influenced by an electrogenic component of the (Na,K)-pump. As illustrated in figure 1, K efflux, measured in rat renal proximal tubular cells (Crabos et al., 1988) and in rat hepatocytes (Cohen and Lechene, 1988), was independent of (Na,K)-pump activity over a three fold range of pump rate.

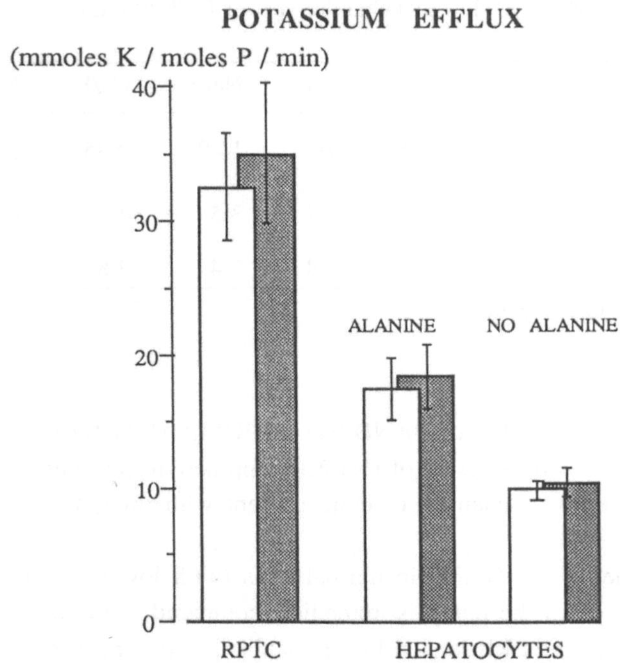

Figure 1. Initial rate of K efflux was measured with the (Na,K)-pump active, using Rb substitution for K in the experimental medium or after (Na,K)-pump inhibition with 1mM ouabain. Wistar Kyoto rats renal proximal tubular cells were in Dulbecco Eagle modified medium. Rat hepatocytes were in Ringer without or with alanine (10 mM). Open bars: Rb substitution; Dark bars: 1mM ouabain.

Na-K pump activity can be measured in three conditions.

1. (Na,K)-pump at Vmax activity. The cells are loaded with Na using preincubation in K-free medium. At the start of an experiment, the cells are quickly switched in 5mM K-containing medium. The ouabain sensitive initial rates of Na efflux and of K influx is taken as a good estimate of the Na-K

pump activity (Harris et al., 1986 ; Larsson et al., 1988). The measurement reflects the amount of transport protein present in the cell membrane.

2. (Na,K)-pump activity in control, steady state situation. The cells contain a normal, low intracellular Na content. Cells are preloaded with Rb by incubation in K-free medium containing 5mM ultra pure RbCl; in this situation the cells exchange all intracellular K for Rb. After the cells are returned in a medium containing 5mM KCl and Rb-free, the ouabain sensitive rate of K influx is taken as the rate of steady state (Na,K)-pump activity for K (Harris et al., 1986).

3. (Na,K)-pump rate at different intracellular concentration of Na. After cells are preloaded with Rb, they are incubated in a medium lacking both K and Rb, for variable periods of time. The (Na,K)-pump is then inhibited and cell load Na in amount variable with the duration of pump inhibition. At the start of an experiment the cells are returned in a medium containing 5 mM K. Initial rate of ouabain sensitive K uptake is taken as an estimate of (Na,K)-pump activity, at the value of intracellular Na measured at the start of an experiment. A series of such measurements on aliquots of the same cells preincubated for variable time, thus with different intracellular Na, allows us to estimate an apparent affinity of the (Na,K)-pump for Na (more precisely for intracellular Na + K) (Crabos, Cantiello, Sweadner and Lechene,1988 ; Crabos Cantiello, Bianchi and Lechene, 1988).

II. APPLICATIONS

Electron probe analysis of cultured cells has the major advantages to allow one to study not only steady states concentrations but to characterize the components of ionic leak pathways, the determinants of (Na,K)-pump activity and the interplay of K, Na, Cl and H (in association with fluorescence microscopy). Studies may be performed not only in established cell lines, but also in very short duration primary culture of cells isolated from highly differentiated organs. Such studies underscore the central role of the (Na,K)-pump in mammalian cellular life (Lechene, 1988).

Using electron probe analysis of cultured cells we have demonstrated that :

a. in renal proximal tubular cells :

1. 85% of sodium entry is through Na-H exchange (Harris et al., 1986);

2. after cellular acid load, there is an immediate increase in Na entry, followed within seconds by an increase in (Na,K)-pump activity (Harris et al., 1986);

3. after only 4 days of culture, cells return to the level of leak and (Na,K)-pump activity characteristic of the non terminally differentiated state that we observed in immature kidneys (Larsson S. et al., 1986 and 1988);

4. during terminal differentiation increased expression of Na-H exchange precede (and may be the signal) for increased (Na,K)-pump activity (Rane et al., in press);

5. volume regulatory decrease is due to increased K efflux;

6. volume regulatory increase is due to both, a decrease in K efflux and an increase in (Na,K)-pump activity (Harris et al., submitted for publication ; Lechene, 1988);

7. in Milano hypertensive rat, the Km of the (Na,K)-pump for Na is higher than in the normal strain, but the apparent Vmax are equivalent (Crabos, Cantiello, Bianchi and Lechene, 1988);

b. in rat inner medullary collecting duct cells, there is an Na channel inhibitable by amiloride with a K_i of 2.21×10^{-7} M (Blumenfeld et al., 1988);

c. in rat hepatocytes, Na-alanine cotransport increase Na influx and steady state (Na,K)-pump activity three folds. Na-alanine cotransport increase K efflux even when the (Na,K)-pump is inhibited(Figure 1) (Cohen and Lechene, 1988);

d. in coculture of MDCK ouabain sensitive and ouabain resistant cells, protection of the ouabain sensitive type by the ouabain resistant is characterized by the maintenance of normal intracellular high K and low Na, despite the lack of electrical coupling between the cocultured cells(Bolivar et al., 1987)

e. in shark rectal gland cells, steady state Na content decreases when the gland is stimulated to secrete(Figure 2);

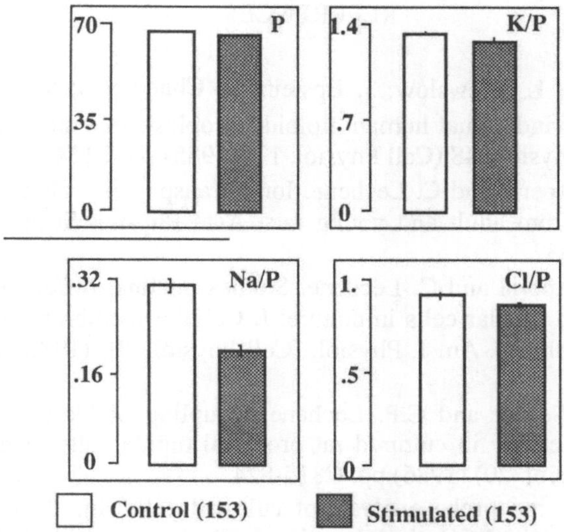

Figure 2. Shark rectal gland cells. Steady state content was measured in unstimulated cells (white bar) and in cells stimulated with 1mM dibutiryl cyclic AMP, 1 µM VIP and .5 mM Theophylline (shaded bar). P : Femtomoles/100 μm^2 area of dried cell analysed. K/P, Na/P , Cl/P : non dimensional units, equivalent to mole element/mole phosphorus. (F.H. EPSTEIN, B.J. COHEN, AND C. LECHENE, unpublished observations).

f. in CV-1 green monkey kidney cells transfected with an ouabain resistance gene, there is ouabain inducible and reversible expression of an ouabain resistant (Na,K)-pump. When the cells have become ouabain resistant, the (Na,K)-pump becomes highly amiloride sensitive (Epstein and Lechene, 1988)

g. in a human cell line fibroblasts, the concentration of polyanions non-diffusible charges is in the 100 mEq range, and is pH sensitive with an iso-electric point around 4.2 (Both and Lechene, 1988).

REFERENCES

Abraham, E.H., J.L. Brewslow, J. Epstein, P. Chang-Sing and C. Lechene. Preparation of individual human diploid fibroplasts and study of ion transport. Am. J. Physiol 248 (Cell Physiol. 17) (1985) pp. C154-C164.

Larsson, L., A. Aperia and C. Lechene. Ionic transport in individual renal epithelial cells from adult and young rats. Acta Physiol. Scand. (1986) vol. 126, pp. 321-332.

Larsson, S., A. Aperia and C. Lechene. Studies on final differentiation of rat renal proximal tubular cells in culture: I. Cellular membrane Na and K effective permeability. Am J. Physiol. (Cell Physiol. 20) (1986) vol. 251, pp. C455-C464.

Harris, R., J.L. Seifter and C.P. Lechene. Coupling of Na/H exchange and Na-K pump activity in cultured rat proximal tubule cells. Am. J. Physiol. 251 (Cell Physiol. 20) (1986) pp. C815-824.

Lechene, C. Electron probe analysis of cultured cells. In: Recent Advances in Electron and Light Optical Imaging in Biology and Medicine, Andrew P. Somlyo, ed. Annals of New York Academy of Sciences, New York, 1986, pp. 270-283.

Bolivar, J., A. Lazaro, S. Fernandez, E. Stefani, V. Pena Cruz, C. Lechene and M. Cereijido. Rescue of a wild-type MDCK cell by ouabain-resistant mutant. Am. J. Physiol. (Cell Physiol.) (1987) vol. 22, no. 1, pp. C151-C161.

Larsson, S.H., A. Aperia and C. Lechene. Studies on terminal differentiation of rat renal proximal tubular cells in culture: ouabain-sensitive K and Na transport. Acta Physiol. Scand. (1988) vol. 132, pp. 129-134.

Epstein, J.A. and C. Lechene. Ouabain-resistant, amiloride-sensitive Na^+-K^+ pumping activity and morphological changes are inducible. Am. J. Physiol. (Cell Physiol. 23) (1988) vol. 254, pp. C847-C854.

Lechene, C. Physiological role of the Na-K pump. In: The Na^+,K^+-Pump: Part B: Cellular Aspects. Proceedings of the 5th International Conference on Na^+,K^+-ATPase. J.C. Skou, J.G. Nørby, A.B. Maunsbach, M. Esmann, eds. Arthur R. Liss, Inc., New York, 1988, pp. 171-194.

Harris, R.H., V. Savin and C. Lechene. Changes in ionic content of rat proximal tubule cells in media of varying osmolality. (submitted).

Rane, S., Y. Fukuda, S. Larsson, A. Aperia and C. Lechene. Developmental increase of Na influx precedes that of (Na,K)-ATPase activity in rat renal proximal tubular cells. (in press,Am. J. Physiol.).

Cohen, B.J. and C. Lechene. Alanine increases Na and K leaks in cultured rat hepatocytes independently of Na-K pump activity. FASEB J. (1988) vol. 2 (5), p. A1490.

Blumenfeld, J.D., H.F. Cantiello and C. Lechene. Ion transport properties of rat inner medullary collecting duct cells (IMCD) in primary culture. 42nd Ann. Meeting of the Soc. of Gen. Physiol. (1988).

Both, G. and C. Lechene. Effects of alanine on ionic fluxes and content in cultured rat hepatocytes. 42nd Ann. Meeting of the Soc. of Gen. Physiol. (1988).

Crabos, M., H.F. Cantiello, K. Sweadner and C. Lechene. Na-K pump activity in normotensive (WKY) and spontaneously hypertensive (SH) rat renal proximal tubular cells (RPTC). 42nd Ann. Meeting of the Soc. of Gen. Physiol. (1988).

Crabos, M., H.F. Cantiello, G. Bianchi and C. Lechene. Differences in intracellular ionic regulation of proximal tubular cells cultured from normotensive (MNS) and hypertensive (MHS) Milan-strain rats. 21st Ann. Meeting of the Am. Soc. of Nephrol. (1988).

ACKNOWLEDGEMENT.

This work was supported by NIH grant RR-02604

Conference Discussion

A. Dörge: You mentioned experiments in which the uptake of K was determined in cells preloaded with Rb. Can you get from these experiments any indication for Rb recirculation?

C. Lechène: In our experimental conditions, initial rates of Rb leak (in Rb loaded cells) are equivalent to initial rates of K leak in primary cultures of rat renal proximal tubular cells, inner medullary collecting duct cells and rat hepatocytes; as well, ouabain inhibitable K or Rb uptakes are equivalent. There is no measurable recirculation of Rb leaking from the cells and repumped in. We have, however, indication of recirculation in the following experiments: the initial rate of leak of K that we measured is higher after placing the cells in a medium without K and with ouabain than after placing the cells in a medium only without K. We interprete this observation as indicating repumping of the K leaking from the cell by the (Na,K)-pump when the cells are in a medium without nominal K but without additional (Na,K)-pump inhibition by ouabain.

K. Zierold: In my experience washing of epithelial cells by water before cryofixation affects the intracellular ion composition whereas in fibroblasts the K/Na ratio seems to be maintained. Can you comment on the sensitivity of the intracellular ion content as depending on the composition of the washing medium for different cells?

C. Lechène: In any of the cell types we have or are studying, from connective epithelial, or blood tissues: NIH 3T3 fibroblasts, Hras and Kras transformed NIH 3T3 fibroblasts, EC10D human fibroblasts, human diploid fibroblasts, mesangial cells, bovine aorta endothelial cells, bovine retinal endothelial cells, primary epithelial culture of rat gastric surface cells, rat renal proximal tubular and inner medullary collecting duct cells, and epithelial cell lines: MDCK, LLC-PK, and CV1, distilled water was found to be the washing media of choice. The absolute K content is maintained at values equivalent to the values found when cells are washed in isosmotic NaCl or ammonium acetate. The K/Na ratio is either equivalent or higher than when washed in isotonic ammonium acetate. Our washing conditions are rigid: 6 sec total in ice cold media immediately followed by quenching. Our samples are very small (2x2 mm). Dr. John Cook, however, told us that he has used distilled water wash on epithelial LLC-PK_1 cells cultivated on large Petri dishes with success - Epithelial cells are more leaky than fibroblasts (half time of K efflux is approximately 100 min for 3T3 fibroblasts and 15 min for rat renal proximal tubular cells). We maintain the cells under incubator conditions (37^oC, gassed with air and 5% CO_2) until the very moment of washing.

Quantitative X-Ray Elemental Mapping of Dynamic Physiologic Events in Skeletal Muscle

Peter Ingram*, Rashid Nassar+, Ann LeFurgey+, Scott Davilla**, and Joachim R. Sommer#

* Dept. of Pathol., Duke Univ.Med.Ctr.
and Research Triangle Institute
Research Triangle, N.C. 27709 USA

A striking event in animal biology, both fundamental and universal, is the process of coupling excitation to contraction (ECC) in muscle. The first step in that process, the excitatory stimulus, is electrical and is followed by the release of free calcium from known intracellular stores called the "junctional SR" (JSR, Sommer, 1968). The freed calcium unlocks a cocked state of an actomyosin interaction at rest, causing the muscle to contract. Neither the mechanism by which the electrical signal is translated into calcium release, nor its precise time course are known. Whereas our initial aim has been to define the accurate time of that calcium release in single, intact frog skeletal muscle fibers (*R. temporaria*), our overall goal is the description of the quantitative changes in elemental microtopochemistry during the entire time-course of excitation-contraction coupling, in both skeletal and cardiac muscle. By measuring calcium directly, we are in a position to avoid the complexities introduced by as yet poorly-understood interactions between calcium and a number of indicator substances (Blinks, 1978; Baylor, 1988) that result in their emitting measurable calcium-dependent signals. Given the present model, we also have the unique opportunity to explore the limits, *per se*, of electron probe x-ray microanalysis (EPXMA) as a legitimate means for the study of dynamic physiologic events dominated by spatial displacements of elements. Single intact cells, excellent time resolution and optimal cryotechniques are imperative for such a purpose (Nassar, 1986; Sommer, 1988). Skeletal muscle fibers were used because they have many very large JSRs containing very high concentrations of calcium (Somlyo, 1980).

The whole time-course of ECC in muscle has two major components: 1) excitation-calcium release (ECR) and, 2) the contraction-relaxation cycle proper (CRC). In all, ECC takes about 200 ms (depending on temperature and type of muscle) from stimulus to relaxation. So far, we have only addressed the first component of ECC, i.e. the time-course of calcium release following electrical stimulation (ECR) when suddenly large

---

\+ Div. of Physiol., Dept. of Cell Biol., Duke Univ. Med. Ctr.
\# Dept. of Pathol., Duke Univ. Med. Ctr.; Durham, NC 27710
** Dept. of Cell Biol., Duke Univ. Med. Ctr.; Durham, NC 27710

quantities of calcium are moved within a short time from a point of very high to one of very low concentration. Whereas, given favourable statistics, the time-course of calcium release can be plotted by measuring concentration changes in time, the *absolute* calcium concentrations are of less significance within this context. Below we show our early results using EPXMA in the form of quantitative elemental mapping of large regions of freeze-dried cryosections of single, intact frog skeletal muscle fibers in an attempt to quantify time-dependent, *quasi in vivo* physiologic events, i.e. the displacement of calcium from sharply defined stores of high concentration (the JSR) after electrical stimulation and, on a slower time scale, after a pharmacologic intervention (ryanodine, Jenden, 1969) that presumably acts on the calcium release channel (Fairhurst, 1970, Fleischer, 1985, Pessah, 1985, Lai, 1988). The former addresses, directly, a fundamental question of striated muscle function, namely the time interval between electrical stimulation and activation of the calcium release channels. The semi-quantitative ryanodine experiment, on the other hand, highlights the unique potential of EPXMA as a method of enquiry, which is to confirm or reject, at the ultrastructural level for the *quasi in vivo* case, such conclusions concerning cell function as may be suggested by biochemical experiments performed at the molecular level *in vitro*.

MATERIALS AND METHODS

The methodology consists of two parts: 1. Tissue preparation and, 2. EPXMA. Tissue preparation and x-ray microanalysis were performed in different laboratories, physically separated by about half a kilometer: no specimen deterioration was observed on transit.

Our approach has been to quick-freeze isolated, single, intact frog skeletal muscle fibers at known time intervals after electrical field stimulation, and to analyze thin sections of freeze-substituted, and carbon/platinum replicas of freeze-etched preparations from these fibers by EM, and adjacent cryosections by quantitative elemental EPXMA (Figs.1-11), all from the same fiber.

Tissue preparation

The methods of Nassar (1986) were used. Briefly:

Tissue: Single, intact muscle fibers from the semitendinosus muscle of *R. temporaria* were isolated and transferred in frog Ringer's solution to a specimen holder that fits a device (Cryopress, MedVac) which allows the muscle fibers to be dropped on command onto a polished, liquid-He-cooled copper block (Heuser, 1976, van Harreveld, 1964).

Timing: At an appropriate moment during the descent of the specimen toward the cold copper block, electronic circuitry (Nassar, 1986) permitted an electrical stimulus simultaneously to excite the fiber and to start a digital clock. The clock was stopped by a circuit activated upon impact of the fiber on the copper block. Stimulation voltage was 1.5 times threshold.

Transfers: After sectioning (Reichert Ultracut E4 Ultramicrotome with a FC-4 attachment) and external transfer, the cryosections were freeze-dried in a Balzers FDU 010 freeze-drying unit with a turbomolecular vacuum pump. The ambient vacuum was monitored with an AMETEK residual gas analyzer. The freeze-drying proceeded against a liquid-nitrogen cold trap (14 mm above the specimens) at a warming rate of 0.1 ^{o}C/min starting at about liquid nitrogen temperature. The vacuum in the chamber during freeze-drying varied between $5x10^{-6}$ to 10^{-7} torr. The freeze-dried cryosections were carbon-coated at 35 ^{o}C without breaking the vacuum; the thickness of the coats was controlled by a Balzers quartz monitor. Thereafter, the sections were transferred externally into plastic grid boxes (Ernest Fullam, Schenectady, NY, green) and stored in a vacuum desiccator attached to an oil-free membrane pump at room temperature. All connectors, containers and tubing were examined by EPXMA for absence of potential contaminants, especially Si, Na, P, Cl, Ca, K and S. Sections have been kept in these enclosures for many months without observable changes in elemental composition.

Cryopreservation: Criteria for early cryodamage established previously (Nassar, 1986) were applied to thin sections of Epon-embedded freeze-substituted and freeze-fracture replicas (not shown) of parts of the same muscle fibers (cf.Figs.1,4), and compared with the appearance of adjacent freeze-dried cryosections. Cryoartifacts were absent in the thin sections of freeze-substituted preparations for about 2 μm into the tissue (Figs.4,11). The "Z-line criterion" (i.e. smooth, fuzzy versus streaky Z lines, Nassar, 1986), was discernible in some thin cryosections. Judging by such criteria, the cracks in the section of Fig.1 are most likely due not to cryodamage but possibly to shearing forces from sectioning, as shown by the smooth Z lines within the first μm from the cell surface in both the cryosection (Fig.1) and the adjacent freeze-substitution preparation(Fig.4).

Ryanodine experiments: Single, intact skeletal muscle fibers were isolated and exposed to $2x10^{-4}$M Ryanodine (courtesy Dr. W. Hasselbach) in frog Ringer's solution for 1, 2 and 3 minutes, respectively, while being stimulated at 2 Hz. After 2 minutes, the contractile response became sluggish. The fiber at 3 min was quick-frozen just prior to irreversible contracture. A quantitative elemental x-ray map was prepared from one fiber at each time point, and small raster probes of at least 4 JSRs performed, and the average amount of calcium was expressed in mmol/kg dry weight.

Electron probe x-ray microanalysis

Freeze-dried cryosections were analyzed in a transmission electron microscope (JEOL 1200EX TEMSCAN) equipped with a scanning device, an additional hard x-ray aperture, a 30 mm^2 Si(Li) detector and multi-channel analyser (Tracor Northern 5500), and a liquid-nitrogen-cooled, low-background cryotransfer stage (Gatan 626). X-ray spectra were obtained for 500 s per area from selected cell regions (e.g. JSRs, A bands, etc.) at approx. $-110^{\circ}C$ by use of a small, square raster (0.06 - 0.1 μm) at 100,000 to 150,000X magnification with specimen tilt at 35°, 80 kV accelerating voltage, and beam current of approx. 0.1 nA. Techniques for obtaining quantitative elemental x-ray images were employed as previously described (Ingram, 1988). Briefly, 64x64 or 128x128 pixel images were obtained at 15,000 to 30,000 magnification (1 pixel approx. 0.065 - 0.13 μm) for physiologically relevant ions (Na, Mg, P, S, K, Cl, Ca) using a beam current of approx. 1 nA and dwell times of 1 or 2 s per pixel. All maps were obtained at ambient temperature (approx. $23^{\circ}C$) with no evident specimen contamination. Digital image acquisition and spectral processing were performed with a multi-tasking graphics workstation (VAXstation 2000, Digital Equipment Corp.); this enabled simultaneous display in real time of separate processes, such as the raw and filtered spectra, spectral peak values vs. livetime and STEM intensity line profiles, etc. For ease of use and reasons of low cost, data storage, image display, processing and retrieval of quantitative image information were performed with a microcomputer (Macintosh II, Apple Computer). The microcomputer user-interface program reads and displays the saved floating-point x-ray images. The mouse function serves to draw boxes, circles and arbitrary polygons on one image map, transferrable to all others. Complex statistics relating to any region of the image maps are displayed on the screen (within 1 s for a 64x64 pixel matrix, depending on the complexity of the region, Ingram, 1988).

It should be noted that the microcomputer itself can be used as both the x-ray analyzer (Fiori, 1988), spectrum processor and beam positioner (Kopf, 1988); the latter greatly simplifies interactions such as on-line drift correction (Statham, 1987) and programmed static probing. We correct for drift *post facto* using a general-purpose image-correlation program (Ingram, 1987).

All spectral data from probes and maps were processed for quantitative analysis by the Hall continuum normalization method (Hall, 1979) with the peak centroid shift and broadening corrections of Kitazawa and collegues (Kitazawa, 1983) and appropriate correction for the carbon support film thickness. Standards were prepared according to the methods reported earlier (Shuman, 1976). Typically, quantitative data were retrieved from the maps by placing a 3 x 3 matrix of pixels or a single pixel over the area of interest and reading the values directly from the computer screen.

Potential mass changes in the specimen due to the electron beam or to shift during probing were monitored every 10 s. The microscope and stage were baked out periodically; the stage was left in the instrument when not in use at ≈ 70 °C to minimize residual contamination in the region of the specimen; the vacuum was monitored with a residual gas analyzer (AMETEK, Inc.). Following initial mass loss at the high beam currents used, further mass changes were always less than 1% and were assumed to be the same for standards and sample.

RESULTS AND DISCUSSION

Quantitative digital elemental imaging (mapping) from electron probe x-ray microanalysis has been demonstrated to be a powerful tool for addressing problems in cell physiology (LeFurgey, 1988, Somlyo, 1984, Johnson, 1988, Saubermann, 1988). In addition to time saved in data acquisition, an essential degree of overall perception is added (Ingram, 1988). While at present mapping is not as accurate as static probes acquired for long periods of time from specific regions of cryosections, it permits reasonably good quantitative data to be obtained from a much larger number of regions within the same or different cells, and meaningful biological trends can be established in the same time it would take to do just a few static probes.

Our data from relatively low-statistic quantitative elemental maps confirm the earlier findings with static probes by Somlyo and colleagues (Somlyo, 1980) in skeletal muscle JSRs at rest and in tetanus (mean [Ca] ± SD mmol/kg dry wt, rest: 103 ± 34 (n=246), tetanus: 55 ± 17 (n=53); $p < .001$). Our data also show a significant difference ($p < .001$) between JSR mean [Ca] at rest (103 ± 34; n=246) and 1.8 - 10 ms after stimulation (83 ± 30; n=398). Whereas it is not yet possible to present the time-course of calcium displacement during ECR (e.g. at approx. 0.1 ms intervals) with a comfortable degree of statistical significance, the methodology presented here appears quite capable of accomplishing the feat. Having established the efficacy and reliability of the biological model (Nassar, 1986) and the data collection (Ingram, 1988), we are now pursuing, judiciously and simultaneously, several options to raise the levels of statistical significance with the view to optimizing our time commitment. At the level of data collection, for example, when the dwell time was increased from 1 to 2 s (which increased mapping time from 2 to 4 hr), the measured elemental concentrations remained identical whereas the variance was reduced. Moreover, quantitative scanning maps at twice the resolution, albeit at quadruple the mapping time, provided better accuracy of data collection from the computer screen (Figs.5,6) because overlaps (due to pixel geometry) of adjacent JSR and

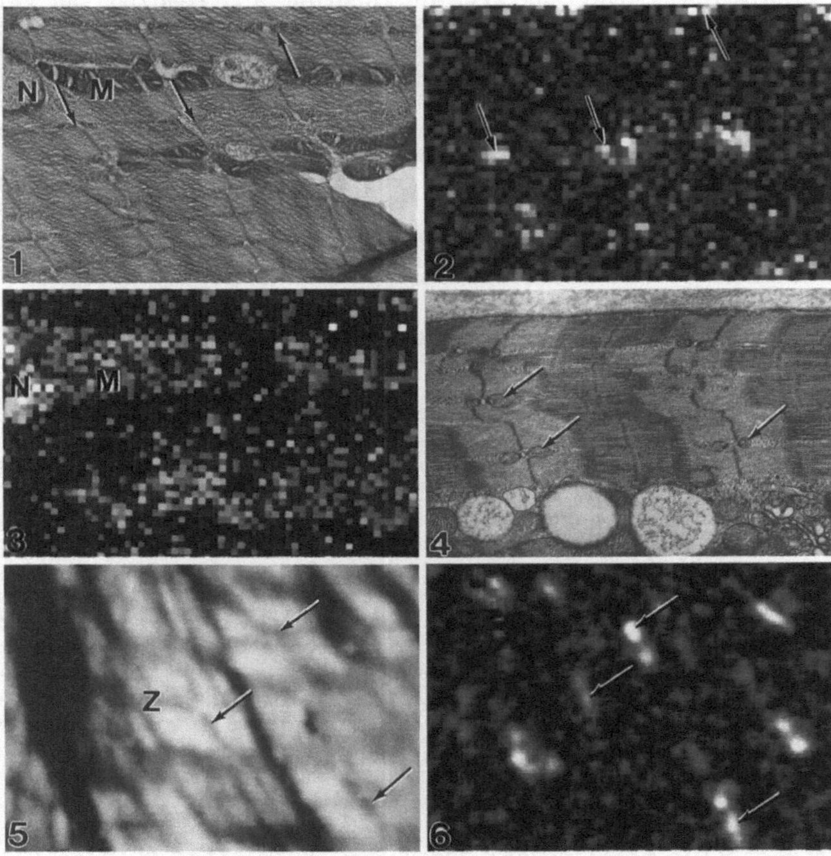

Fig.1 Ryanodine (1 min). Unstained cryosection. JSR of triads = arrows. Mitochondria = M. Nucleus = N. 9,000X

Fig.2 Quantitative x-ray [Ca] map (64/64 pixels) showing correspondence of high [Ca] with JSRs (arrows) in Fig.1. Color scale: highest [Ca] white, lowest black.

Fig.3 Quantitative x-ray [P] map showing correspondence of high [P] with mitochondria (M), the nucleus (N) and, less so, JSRs of Fig.1.

Fig.4 Thin section of freeze-substituted muscle adjacent to cryosection of Fig.1. JSRs=arrows. Mitochon.= M. 14,000X

Fig.5 256x256 pixel STEM image of different section from same fiber as Fig.1 taken under analytical conditions (large spot size, ~10nA probe current). JSR=arrows, Z=Z line.

Fig.6 High resolution (128x128 pixels) quantitative calcium map corresponding to Fig.5. JSR = arrows.

non-JSR regions could be reduced; indeed, quantitative data from single pixels chosen for the highest [Ca] over JSR regions, more closely paralleled the higher values seen, on the average, in static probes. The employment of a field emission gun is an additional important option for improving statistics. Finally, statistics can be improved at the level

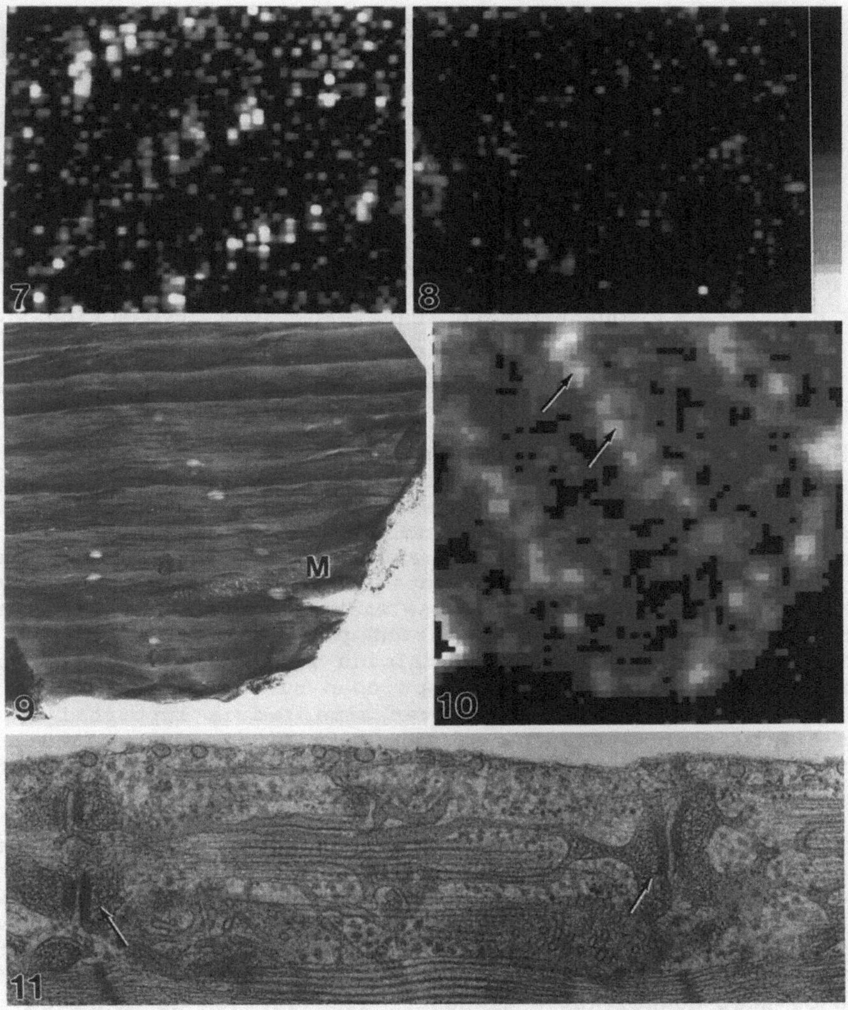

Fig.7 Quantitative x-ray [Ca] map of a muscle fiber at rest. Color scale: highest concentration = white.
Fig.8 Quantitative x-ray [Ca] map of muscle in tetanus. Color scale as in Fig.7.
Fig.9 Unstained cryosection, TEM image. Note poor structure. Mitochondrion = M. 16,000X
Fig.10 Quantitative x-ray [Ca] map of cryosection in Fig.9. Discrete localization of high [Ca] in presumed JSR locations (arrows).
Fig.11 Thin section of freeze-substituted fiber region adjacent to cryosection of Fig.9. JSR = arrows. 46,000X

of the preparation: it is possible to cut cryosections down the length of a quick-frozen single muscle fiber at successive

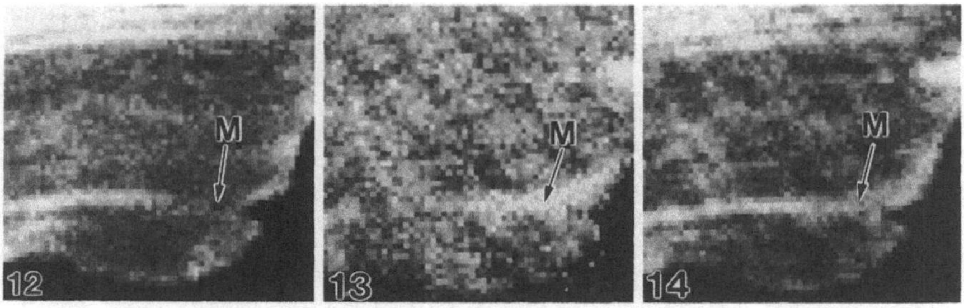

Figs. 12 - 14 Quantitative image maps from cryosection shown in Fig.10. In the [P] map (Fig.13) the mitochondrion (cf.Fig.9) with its high P content is highlighted, in contrast to the continuum map (Fig.12) which reflects overall mass effects, and the peak P map (Fig.14) which reflects both. M = mitochondrion.

intervals away from a single point of electrical stimulation. By taking advantage of the finite propagation velocity of the excitatory wave, elemental displacements at the surface can be put on a time scale in a single fiber, thus eliminating the effects of statistical variance among fibers and among animals.

The ryanodine experiments were especially gratifying because they confirmed *quasi in vivo* what recently has been shown to be the function of the so-called junctional processes of the JSR *in vitro*: they contain the calcium release channels to which ryanodine binds and which, kept open by the poison, drain the JSR of its calcium content over time, as is supported by the present experiments. The average total [Ca] of the JSRs after two min exposure to ryanodine dropped to approx. 50%, and after 3 min to approx. 20% of the control value.

Several aspects of quantitative elemental imaging are apparent. Figs. 1 through 8 show the excellent correspondence between structure and micro-topochemistry. Given high temporal and spatial resolution and good cryopreservation coupled with high resolution quantitative imaging, one might be able, eventually, to visualize directly diffusion from a point in certain regions of cells such as in the present model. Figs. 9, and 10 show that quantitative elemental localization is achieved in a cryosection showing little structural detail as may be the case, for example, in a thicker than usual section, especially if it is free of cryoartifacts, and that might be rejected for that reason for further study. This dramatic localization indicates: 1) that the appearance of well-defined "morphology" in a conventional TEM or STEM image is not *a priori* a necessary criterion for obtaining quantitative elemental distributions in cryosections and, 2) while not permitting any conclusions whatever concerning the quality of tissue preservation *per se* (e.g. alterations occurring during freeze-drying or transfer of cryosections must be assessed independently, even as the adjacent freeze-substituted section, Fig. 11, shows no cryoartifacts in this particular case), some distinct micro-topochemical information may be disclosed, e.g. the property of the JSR to sequester calcium and to retain it locally, even in adversity (the K/Na was low in this section). Conversely, prominence of structural detail, e.g. the degree of electron-density of JSRs or other structures in the STEM image, may be unrelated to actual elemental concentrations. Quantitative elemental maps are also uniquely suited to demonstrate unambiguously that the Hall continuum normalization procedure (Hall, 1979) is clearly effective in thin cryosections (Figs.12,13, 14). Several horizontal streaks of contrast in the section (folds, cf. Fig.9) seen in the continuum (Fig. 12) and in the P peak map (Fig.14) are totally eliminated (normalized) from the P concentration map (Fig.13). Finally, quantitative elemental maps, recorded permanently, also offer unlimited future access for additional numerical exploitation unanticipated at the time of data acquisition. Color-enhancement of quantitative elemental maps, especially if they are of high resolution, will aid substantially in revealing differences in [Ca] within the JSR itself, for example, between the electron-lucent core cylinder (Nassar, 1986) which may function as a calcium pool ready for instant export, and other parts of that organelle.

Acknowledgments:

The superb technical accomplishments of Teresa High, Elsa Scherer and Isaiah Taylor at many levels in support of this research is both admired and fully appreciated.

This work has been presented, in part, at the 46th Annual Meeting of the Electron Microscopy Society of America, and the 23rd Annual Meeting of the Microbeam Analysis Society. Milwaukee, WI, 8 - 12 August, 1988.

This work was supported by NIH grant # HL-12486 and the Veterans Administration Research Service.

REFERENCES

Blinks JR, Rüdel R, Taylor RS (1978) Calcium transients in isolated amphibian skeletal muscle fibers: detection with aequorin J.Physiol. 227:291-323

Baylor SM, Hollingsworth, S (1987) Fura-2 calcium transients in frog skeletal muscle fibers. J. Physiol. 403:151-192

Fiori C (1988) Some NIH programs for image analysis, stereology and multichannel analysis for the Macintosh. Joint meeting of the 46th EMSA and 23rd MAS, Milwaukee, WI.

Fairhurst AS, Hasselbach W (1970) Calcium efflux from a heavy sarcotubular fraction. Effects of ryanodine, caffein and magnesium. Eur.J.Biochem. 13:504-509

Fleischer,S., Ogunbunmi,E.M., Dixon,M.K., Fleer,E.A.M. Localization of Ca^{2+} release channels with ryanodine in junctional terminal cisternae of sarcoplasmic reticulum of fast skeletal muscle. Proc.Natl.Acad.Sci.USA 82:7256-7259, 1985

Hall T (1979) Biological x-ray microanalysis. J.Microsc. 117: 145-163
Heuser JE, Reese TS, Landis DMD (1976) Preservation of synaptic structures by rapid freezing Cold Spring Harbor Symp. Quant. Biol. 40:17-24
Ingram P, LeFurgey A, Davilla ., Lamvik MK, Kopf DA, Mandel LJ, Lieberman M (1987) Realtime quantitative elemental analysis and imaging in cells. Analytical Electron Microsc. pp 179-183
Ingram P. LeFurgey A, Davilla SD, Sommer JR, Mandel LJ, Lieberman M, and Herlong JR (1988) Quantitative elemental x-ray imaging of biological cryosections.Microbeam Anal.,p.433-439
Jenden DJ, Fairhurst AS (1969) The pharmacology of ryanodine. Pharmacol. Rev. 21:1-25
Johnson DE, Izutzu K, Cantino M, Wong J (1988) High spatial resolution spectroscopy in the elemental microanalysis and imaging of biological systems. Ultramicroscopy 24:221-236
Kitazawa T., Shuman H., Somlyo AP (1983) Quantitative electron probe analysis:problems and solutions. Ultramicr. 11:251-262
Kopf D (1988) Personal communication
Lai FA, Erickson HP, Rousseau E, Liu Q-Y, Meissner (1988) Purification and reconstitution of the calcium release channel from skeletal muscle. Nature 331:315-319
LeFurgey A, Bond M, Ingram P (1988) Frontiers in electron probe microanalysis: application to cell physiology. Ultramicroscopy 24: 185-219,
Nassar R, Wallace NR, Taylor I, Sommer JR (1986) The quick-freezing of single intact skeletal muscle fibers at known time intervals following electrical stimulation. Scan. Electron Microsc. I:309-328
Pessah IN, Waterhouse AL, Casida JE (1985) The calcium ryanodine receptor complex of skeletal and cardiac muscle. Physiol. Res. Commun. 128:449-456
Saubermann AJ, Heyman RV (1987) Quantitative digital x-ray imaging using frozen hydrated and frozen-dried tissue sections. J. Microsc. 146:169-182
Shuman H, Somlyo AV, Somlyo AP (1976) Quantitative electron probe microanalysis of biological thin sections: methods and validity. Ultramicroscopy 1:317-339
Somlyo AP, Somlyo AV, Gonzalez-Serratos H, Shuman H, McClellan G (1980) The sarcoplasmic reticulum and its composition in resting and in contracting muscle. In: Muscle Contraction: Its Regulatory Mechanisms. Ebashi,S., Maruyama,K., Endo, M. (eds.), p 421. Japan Sci. Soc. Press/Springer Verlag, Tokyo.
Somlyo AP (1984) Compositional mapping in biology: x-rays and electrons. J. Ultrastr. Res 88:135-142
Sommer JR, Johnson EA (1968) Cardiac muscle. A comparative study of Purkinje fibers and ventricular fibers. J.Cell Biol. 36: 497-526
Sommer JR, Johnson EA, Wallace NR, and Nassar R (1988) Cardiac muscle following quick-freezing: Preservation of in vivo ultrastructure and geometry with special emphasis on intercellular clefts in the intact frog heart. J.Mol.Cell. Cardiol. 20:285-302
Statham PJ (1987) Quantitative digital mapping with drift compensation. Analytical Electron Microscopy, pp 187-190
Van Harreveld,A., Crowell,J. (1964) Electron microscopy after rapid freezing on a metal surface and substitution fixation Anat. Record 149:381-386

Conference Discussion

Th. von Zglinicki: My impression is that your frozen-dried sections do not show the same beautiful preservations as the freeze-substituted ones. Is this correct and if so, what are the reasons?
J.R. Sommer: All embedding techniques have been developed to stabilize tissues for smooth sectioning. Cryosectioning is not smooth sectioning and cannot be compared with cutting plastic embedments. Cryosections are always thicker and subject to distortions mostly due to shearing, undamped by the absence of surrounding plastics.

H.K. Hagler: How do you explain the large differences between the ultrastructural preservation preserved by freeze-substitution and that observed in your cryosection, particularly Figs. 5 and 9? In our experience using cryotransfer and freeze-drying in the microscope the ultrastructure should be essentially the same as that preserved by freeze-substitution methods. In comparing the ultrastructural differences between Fig. 1 and Figs. 5 and 9 there is evidence of rehydration of the sections and loss of structural resolution (Hagler, H.K. and Buja, L.M., 1986, J. Microsc. 141:311-317). When working with high resolution microanalysis, SR, mito, etc. it is generally accepted that good morphology is required, otherwise you cannot be certain of either the structure analyzed or the results. Figs. 5 and 9 have no convincing ultrastructure and yet it is claimed that X-ray mapping and image analysis techniques do not require good morphology. How do you rationalize these discrepancies?
J.R. Sommer: To both questions at once: of course, only in some respects (e.g. cryoartifacts) can images from thin sections of plastic-embedded material be expected to be comparable to freeze-dried cryosections cut at 3 to 5 times the thickness. The freeze-substituted thin section of Fig. 11 reveals no perceptible distortions due to ice crystal damage in the adjacent cryosection of Fig. 9. Therefore, this particular cryosection (Fig. 9) is not comparable to any other cryosection that is severely distorted by ice damage. Thus, it is impos-

sible to judge whether the images in question might be related to the presence or absence of such artifacts, or due to other causes, including "rehydration" which remains to be documented by measuring water. At present, alternative explanations must be considered, such as undue thickness of the cryosection and the presence or absence of shearing artifacts that, like ice crystals, create interstices and, thus, contrast between structures (see Fig. 1), adding laudable, if uninterpretable, crispness to the perception of morphology. In the absence of such contrast-creating causes which tend to accentuate underlying structure, a perfectly good section might look quite like a very thick plastic section when viewed with an electron microscope. At this time, we hesitate to ascribe images as seen in Fig. 9 to rehydration, or any other defined artifact over other possible, alternative explanations. We do appreciate the problem as is argued and documented for this particular case (i.e. Fig. 9) in the text of the paper. Clearly, by the definition of micro-topochemistry, and from a practical point of view, only sections with excellent structural preservation are suitable for our study, even as the quality of morphology, however judged, cannot be used without independent confirmation by some other feature, as yet to be discovered, and as a direct, and specific (e.g. "rehydration") gauge of the state of preservation of the original quantitative elemental localization. In the meantime, however, it is important to reemphasize that elemental micro-topochemistry is effective in the absence of visible, detailed structure, if only, as it may turn out, for forensic and archaeologic pursuits, as well as for a full appreciation of how tightly sequestered Ca really is in the JSR, even in adversity.

R. Reichelt: What is the criterion you are using for the correction of specimen drift? It would be of particular interest for a specimen having an irregular structure.

P. Ingram: Specimen drift is corrected post facto by reference to individual STEM images acquired every 8 lines (in the case of a 64x64 pixel image). When the X-ray map is complete, the series of associated STEM images are cross-correlated with an initial STEM image to generate a series of floating point correction vectors. These correction vectors are then applied to appropriate pixels in the X-ray image map resulting in remapping each integer pixel location to a new floating point location. The floating point locations are reversed, bi-linear interpolated to the four integer locations around them, and fractional portions (from the interpolation) of the pixel value are added to the contents at those locations in a new image map. The fraction portions themselves are added to the contents at the same locations in a second temporary map of the same size. After remapping all floating pixel locations back to integer locations, each fractional pixel sum in the remapped image is divided by the corresponding fractional sum in the temporary map.

Since this method (depending upon the rate of drift) might result in empty pixel locations that have no remapped values, a second pass over the new image map is made to interpolate the empty pixel locations from the filed pixel locations around them. This process is simplified by first filling the temporary map with an unique value. In the first pass, when the fractional sum is added to this temporary map, the location is checked for the unique value. If it is present, the fraction sum replaces the value; if not, then the fraction sum is summed with the contents. The second pass again checks for the unique value. If it is present, then the pixel location is empty and the value for that pixel is interpolated from surrounding non-empty pixel locations. This two-pass method undistorts the effect of drift. Additional effects such as sample shrinkage can also be removed by introducing an opposing distortion and then interpolating back to integer pixel boundaries. Obviously, this approach does not depend on the regularity of the structure, since the image is corrected pixel by pixel.

R. Nassar: Could you please comment on your color scheme for quantitative maps, and also whether it can logically be extended to display two elements at once?

P. Ingram: All color schemes for quantitative images of any description should be based on some logical formulation, as was pointed out by K. Heinrich, as well as others, many years ago.

In summary, we use a judicious combination of Hue, Saturation and Intensity (Brightness) extracted from the Chromaticity Diagram (published by the Commission International de l'Eclairage (CIE) in 1976) applied to the particular gray-level image. In the case of Ca we use a scale rather close to the so-called Thermal Scale of black body radiation.

It is certainly possible to display two (or more) elements at once, particularly as ratios which could have some physiological (or physical) meaning eg. K:Na etc. Again, the actual color scale used to display the ratio must have some locigal foundation.

Single Isolated Cardiac Myocytes Frozen During Voltage-Clamp Pulses: A Technique for Correlating X-Ray Microanalysis Data on Calcium Distribution with Calcium Inward Current in the Same Cell

*Maria Fiora Wendt-Gallitelli and +Gerrit Isenberg

INTRODUCTION

Excitation-contraction coupling in heart muscle is mediated by short lasting intracellular calcium transients. This activator calcium stems mainly from intracellular (i.c.) stores (sarcoplasmic reticulum). Extracellular calcium which enters through the sarcolemma via the calcium inward current or the sodium-calcium exchanger is a rather minor source of the activator calcium (Fabiato, 1983). But it is required to fill the i.c. stores and to trigger the release process. Close correlation between increased calcium-concentration due to calcium current, (I_{Ca}) and contractility is demonstrated by a large number of experiments. For example it is well known that application of catecholamines increases I_{Ca} (McDonald and Trautwein, 1978), thus the i.c. stores are filled to a larger extent, (Wendt-Gallitelli, 1986; Wendt-Gallitelli et al., 1982) hence more activator calcium can be released and thereby contraction is potentiated. We have developed a technique which allows to correlate this amount of calcium entering with I_{Ca} with the deriving intracellular calcium distribution. The amount of calcium-entry can be measured by voltage-clamp experiments. But, the intracellular routes of this calcium are absolutely inaccessible by means of electrophysiological methods. Even measurements of the time-course of freely ionized Ca^{2+} with calcium indicators can only incompletely answer the question which fraction of I_{Ca} replenish the i.c. stores, which one modulates the Ca-release and which part serves as activator calcium. But as

+Physiologischees Institut der Universität zu Köln, Robert-Koch-Strasse 39, 5 Köln 41, F.R.G.

*Physiologisches Institut der Universität Tübingen, Gmelinstrasse 5, D-74 Tübingen 1, F.R.G.

far as intracellular calcium is bound or sequestered in small intracellular compartments ("anatomical compartments" limited by a membrane as are sarcoplasmic reticulum, mitochondria, and sarcolemma or "physiological compartments") in relatively high amounts, i.e in the range of at least 100-200 µmol/kg wet weight, it can be measured by x-ray microanalysis (Somlyo et al., 1977; Wheeler-Clark and Tormey, 1986; Wendt-Gallitelli, 1985, 1986; Wendt-Gallitelli et al., 1982).

Increase and decrease of free cytoplasmic calcium at activation and deactivation however, are in the nanomolar range, far below the limit of detectability of the probe. But since this free Ca^{2+} is in equilibrium with a large amount of calcium bound to cytoplasmic soluble and non-soluble ligands, hence during activation of contraction the total Ca-concentration increases up to more than hundred micromolar. Such concentrations are measurable with the probe.

We have now adapted the methods of quick-freezing to ventricular preparations which are subject to voltage-clamp. In order to obtain correct electrophysiological measurements we used isolated ventricular myocytes. Here we present the essential steps of the technique, which can be applied to any kind of isolated cells, whenever the effects of ion-flux controlled by the membrane potential have to be correlated with intracellular elemental-distribution.

The myocytes were isolated from the ventricle of the guinea pig. We used this animal since the contractility of the guinea pig heart strongly depends on the frequency of stimulation, therefore on the history of past excitations over time: contractility decreases to nearly zero when stimulation intervals are prolonged to minutes (rest-contraction) (Wendt-Gallitelli, 1986, Reiter et al., 1984), it increases when the stimulation-interval becomes short, reaching a maximum at paired pulses. One can therefore expect that the calcium-loading of the cell compartments involved in storage of activator calcium reflects the Ca^{2+} increments due to calcium-entry over the time of stimulations as well as the time course of potentiation of contraction.

METHODS

Myocytes were isolated from ventricles of 250 g guinea-pigs by means of the pronase technique (Bendukize et al.1985). The Ca-tolerant myocytes were pipetted into a chamber (2 mm x 2 mm x 10 mm) placed on the stage of an inverted microscope. After the cells had settled at the glass bottom of the chamber, they were continuosly superfused with a prewarmed (35°C) "tyrode solution" of the following composition in mmol/l: NaCl 150, KCl 5.4, $CaCl_2$ 3.6, $MgCl_2$ 1.2, glucose 10, Hepes 10 (adjusted by NaOH to pH 7.4).

The voltage-clamp experiments were performed with a single patch-electrode technique (Isenberg et al., 1987; Marty and Neher, 1983). The electrodes had tips with an outer diameter of 6 µm and an inner diameter of 2-3 µm (resistance 1-1.5 MΩ). The electrodes were filled with the following solution (mmol/l): 140 KCl, 3 Na_2ATP, 3 $MgCl_2$, 0.02 EGTA, 10 HEPES/KOH (pH 7.2). A PDP 11-23+ minicomputer (Digital equipment corp., Maynard, Massachusetts) generated stimuli or clamp pulses (0.5 Hz) and controlled the sequence of pneumatic movements that finally lead to shock-freezing.

Ca-influx was calculated from I_{Ca} which is subject to errors. The recorded current is a net current from which I_{Ca} has to be separated. At present time the rapid solution change inside the taper of the silver holder (see below) was not possible, therefore the "pharmacological dissection" of I_{Ca} was not applicable. Instead, I_{Ca} was approximated with the "visual estimate" which subtracts the net current from a time-independent reference line (taken from the net current as it is measured 0.2 s after start of depolarization (McDonald and Trautwein, 1978)).This method misses the non-inactivating part of I_{Ca} and may underestimate Ca-influx by 20% (0.2 s long pulses) or up to 60% (1 s pulses), respectively (Isenberg and Klöckner, 1982; Isenberg et al., 1987).

In order to calculate from Ca-influx the "apparent increment" the cell volume has to be known. The cell volume was approximated from 1. light microscopy and 2. electrical measurements. Ad 1: The plate area was measured (985 ± 120 µm², n=20) and multiplied with the cell height (10 µm, estimated by focussing the lower and upper surface of the cell). The estimated cell volume was about 10 pl. Ad 2: We measured the cell capacitance (105 ± 13 pF, n=20), divided it by the specific capacitance of 4 µF/cm² (taking into account infoldings etc.) and obtained an outer surface area of 4200 µm². Using a volume to surface ratio of 2.56 µm (rat left ventricle, Moore et al., 1984), multiplication resulted in a cell volume of 10.8 pl. Despite their different source of error, both estimates come to similar numbers.

A patch-pipette served for both the transfer of the cell from the bottom of the chamber to a special silver holder as well as for the electrophysiological measurements. In the hand-made silver holder the tip of a taper, opened electrolytically, was covered with a thin pioloform film. Having selected a myocyte with clear cross-striation the GΩ seal was built but without disrupture of the patch of the membrane.

The cell was lifted from the bottom of the chamber to the holder (Figure 1) and placed on the transparent pioloform film.
During the transfer the cell was exclusively surrounded by the "tyrode"-solution and never came in contact with atmosphere. Intracellular contact

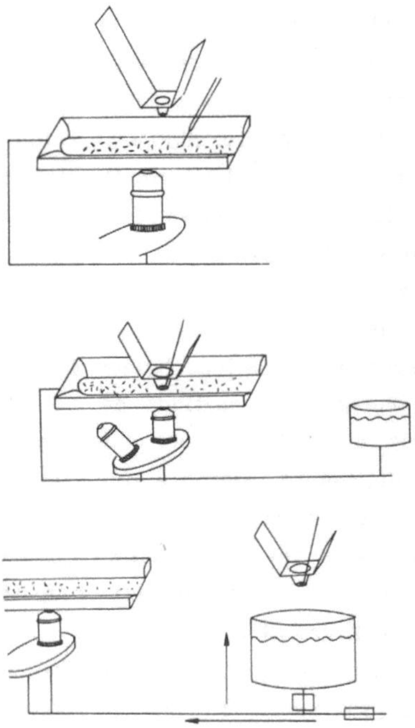

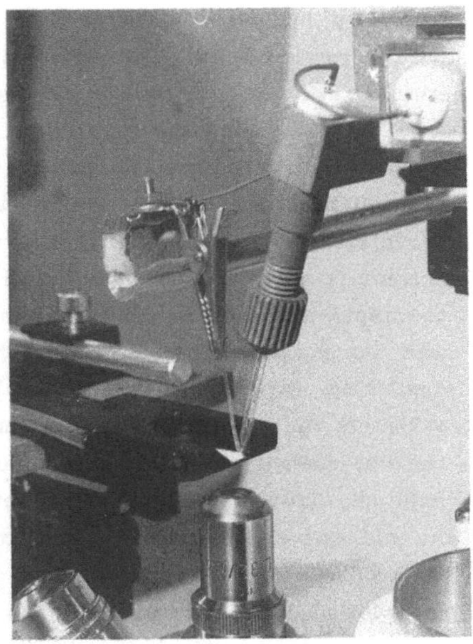

Fig.1: On the left: Schematic of the set-up for transferring and freezing the myocytes. Top: arrangement of microscope, holder, chamber and patch-electrode. Middle: the myocyte has been transferred into the holder. Bottom: microscope and chamber have been moved away, the coolant is moving up to the holder with the cell. On the right: front-view of microelectrode in the holder. The chamber and the microscope-stage are removed away from the holder. In the right corner the container with the coolant is at the same time slided under the holder and then moved up.

was then established by rupturing the patch of the membrane. After checking that the intracellular stimulation induced normal action potentials (Fig 2), voltage clamp was switched on. Pulse depolarizing from −45 to +5 mV evoked net membrane currents which were largerly determined by the Ca inward current. The elecited shortening of the unloaded cells was visualized on the monitor of a TV-microscope.

Then, keeping the holder with the cell absolutely steady, the chamber was moved away. Before the horizontal movement stopped, the vertical movement

>> Slide 2 <<

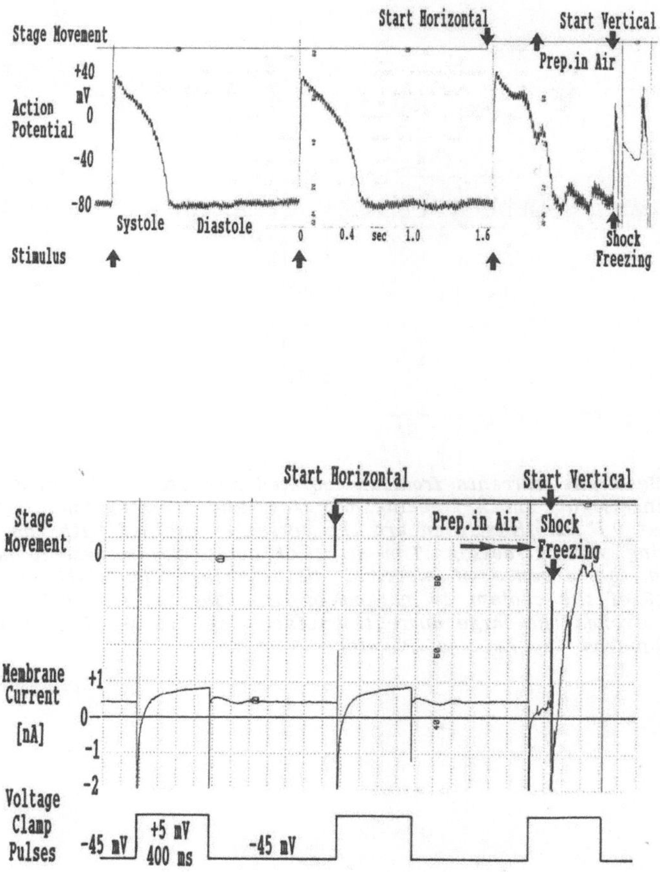

Fig.2: Electrical measurements from ventricular myocytes subject to quick-freezing. Top: Action potentials. Stage movements leading to quick-freezing are labeled. Note: measurements are possible even if sliding away the chamber has positioned the preparation into air. Recording is ended by freezing in diastole. Bottom: membrane Ca-currents (downward deflections) due to voltage-clamp pulses. Shock-freezing is timed during systole (116 ms after start of depolarization).

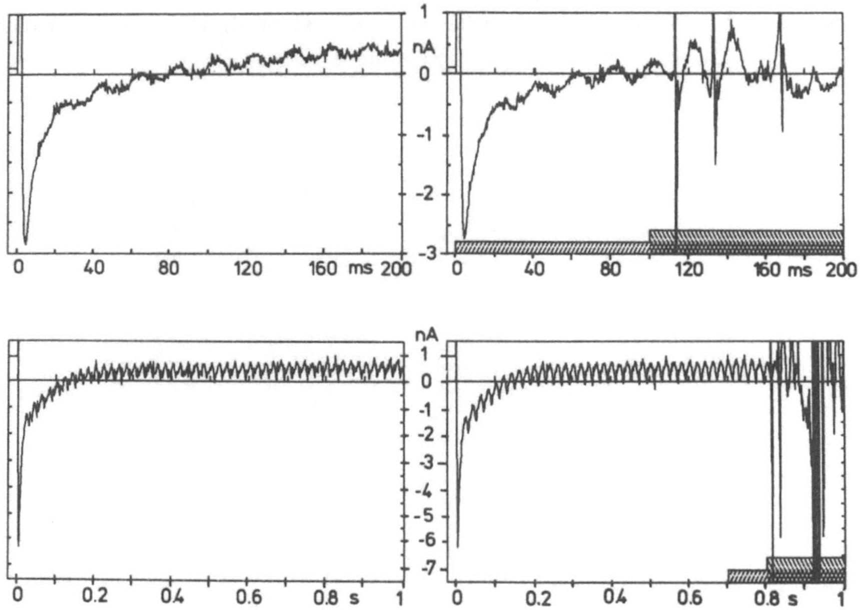

Fig.3: Membrane currents from an isolated myocyte as recorded before (left) and during (right) shock-freezing. 0.2 (top) and 1 s long pulses (bottom) were applied at 0.5 Hz. Illustrated are the currents obtained with the pulse before and during shock-freezing. The start of horizontal movement is marked by ///. Note the absence of elecrical disturbances. The start of the vertical movement of the coolant is marked by \\\. The contact of the coolant with the cell evokes the large current artefact (top 116 ms, bottom 816 ms). (By permission from the Am.J. of Physiol.)

of the coolant started and the preparation contacted the coolant (supercooled propane at −196°C).
The preservation of the cellular structures after freezing is shown in Fig 4. The myocyte was freeze-substituted. Cryoartifacts appeared with a gradient of increasing severity. They were absent at the surface of the cell touching the pioloform film. Cryoartifacts started about 2 µm inside the cell. Most of our shock-frozen cells showed a 2−5 µm thick zone of minimal cryoartifacts. The glass electrode did not limit the quality of freezing.
After freezing the holder is transferred under liquid nitrogen and mounted on the cryoultramicrotome, where ultrathin cryosections are cut at −140°C. In preparing the myocytes for analysis, this step is the most difficult one since we have to cut simultaneously silver, tyrode and muscle, three materials of extremely different ductility. Manipulation of cryosections from a single cell (maximally 150 µm long and 19 µm wide) is demanding, and sometimes small turbulence in the nitrogen-atmosphere of the cryochamber blows away the best cryosections. During cutting it was impossible to visualize the cell as it is embedded in frozen tyrode by means of the normal stereo-microscope (4 x 10), therefore we added a video-system on the basis of a 25fold long-distance Leitz-objective. With the large magnification one can see the myocyte during crycutting, but the reduced visual field makes the manipulations of the grid more difficult.

FIRST RESULTS ABOUT CA-CONCENTRATION IN INTRACELLULAR COMPARTMENTS

In the Siemens Elmiskop the electron beam could be focussed to 50 nm diameter. Using this probe we analyzed the Ca-concentration in compartments as cytosol, mitochondria and junctional SR. For the cytosol the electron beam was located in the space occupied by the overlapping filaments. For the junctional SR the beam was placed on this structure adjacent to the T-tubulus, for the mitochondria the probe was located within these organelles, for the cell borders: the electron beam was deformed astigmatically to a spot of 10 nm x 1 µm, and it was placed as close as possible on to the inner side of the surface membrane. The concentrations sum up from elements bound to the inner side of the sarcolemma, from elements stored in those parts of the junctional SR that were closely attached and elements of the cytosol adjacent to transversal tubuli.

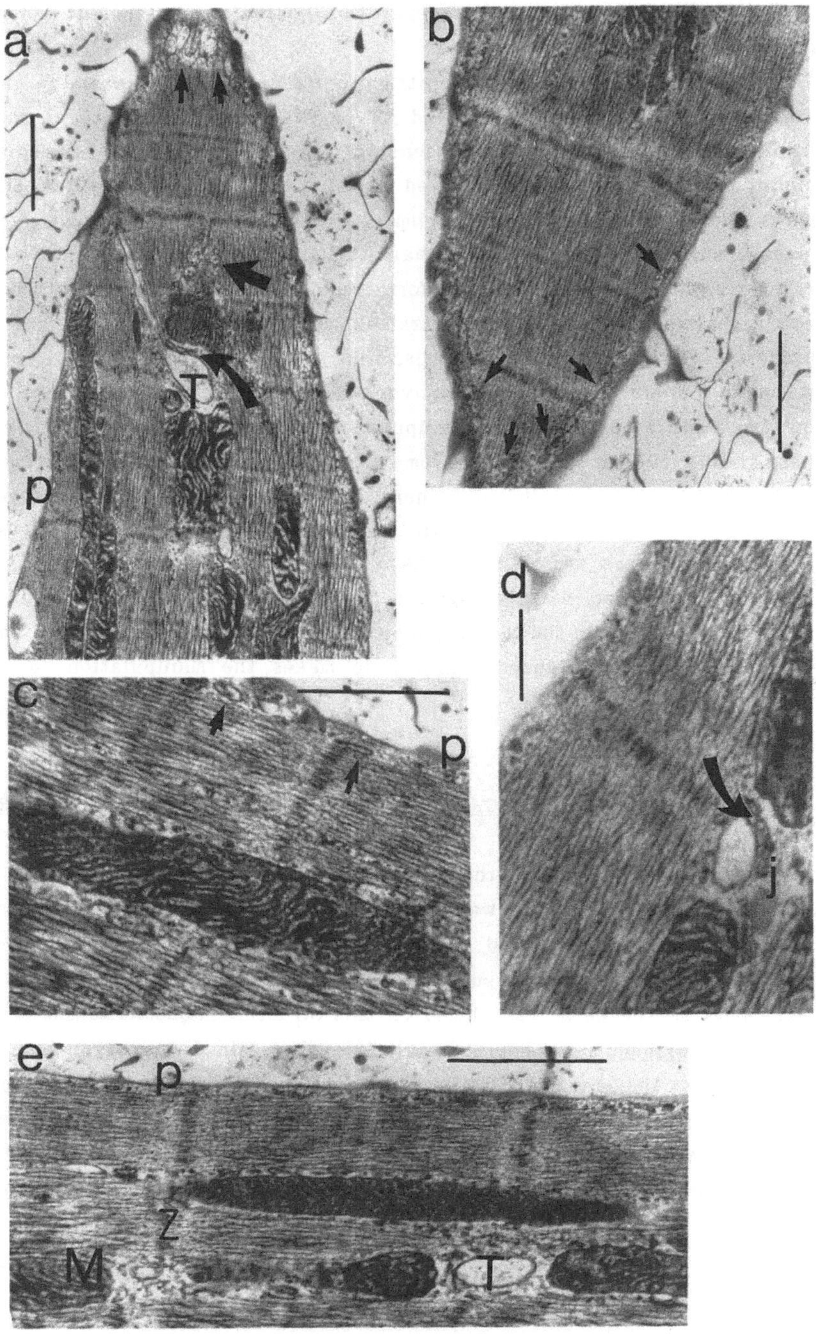
a
p
T
b
c
p
d
j
e
p
Z
M
T

In order to correlate the amount of voltage-controlled Ca-entry to the intracellular Ca-concentration, we performed 3 types of experiments: 1) As a control, the myocytes were frozen after a prolonged rest period (20 min). 2) Following the rest period, Ca-entry (and contractility) were moderately potentiated with a train of 5 pulses of 0.2 s. The cells were frozen 116 ms after start of the fifth pulse, i.e. at the peak of contraction. 3) After rest, strong potentiation was induced by a train of 4 long pulses (1 s duration). The myocytes were frozen 816 ms after start of the fourth pulse, at a time when they had relaxed from contraction.

With this protocol we obtained the following results:

After rest, the Ca-concentration in all compartments was below the limit of detection which is 0.3–0.5 mmol/kg d.w. or ca 75–100 µmol/kg w.w.

Using the moderately potentiated cells, Ca-concentrations were measurable in all compartments (Table 1). But,due to the small number of analysis, the differences between the calcium concentrations in different compartments turned out not to be statistically significant. Strong potentiation with long pulses further increased the Ca-concentration in the compartments, and differences in the calcium-load of the compartments are statistically significant (Table 2).

In the cytosol, the Ca-concentration was 87 µmol/kg w.w. which is far below the Ca-concentration estimated from Ca-entry which amounts to 137 µmol/kg w.w. for the ongoing depolarization and 712 mmol/kg w.w. for all four depolarisations. In light of table 2, this result is thought to suggest that part

---

Fig. 4: *Longitudinal sections of a ventricular myocyte shock-frozen and freeze-substituted. P: plasmalemma, M: mitochondria, j: junctional SR, T: t-tubuli, Z: Z-line.*

a) Minimal cryoartifacts on the left. Some distorsion of the myofilaments on the right which are the first signals of ice crystals damage (cryoartifacts). Bar: 1 µm. Curved arrows: junctional SR, short arrows: subsarcolemmal SR, thick arrows: free network of SR.

b) A better frozen region of the cell. Note on the left the regular pattern of the thick and thin filaments indicating the absence of ice crystals formations during freezing. The arrows indicate to subsarcolemmal vesicles of the SR. Bar: 1 µm

c): Small subsarcolemmal vesicles of the SR are clearly recognizable (arrows). Note the well preserved arrangement of myofilaments and the structure of the Z-lines. Bar: 1 µm.

d) Junctional SR, indicated by the curved arrow, is present at one side of the t-tubulus. Bar: 0.2 µm.

e) Minimal cryoartifacts in the superficial zone indicated by the perfectly parallel arrangement of the filaments and the preservation of the superficial mitochondria. Note the sharp border at the sarcolemma. Bar: 1 µm. (With permission from the Am. J. of Physiology).

>> Slide 10, 11 Kryoschnitte<<

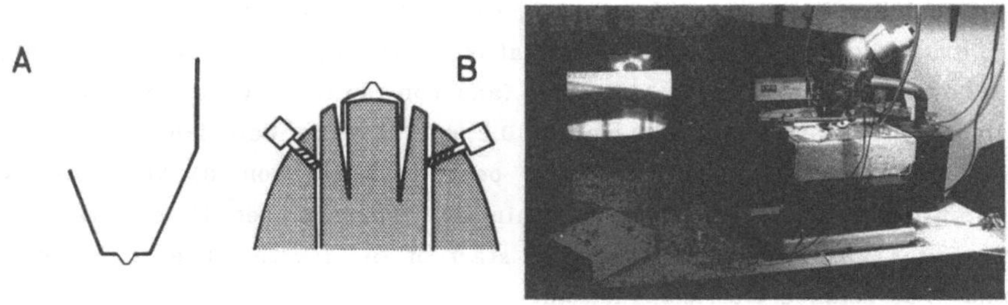

Fig.5: Left: schema of the cell holder (A) and of its adaptation to the LKB cryoultramicrotome (B). Right: Setup used for cryocutting. The TV-monitor shows the knife-edge and the tip of the holder. Distance between 2 cupper bars of the grid (out of focus) is ca 200 nm.

>>cryo<<

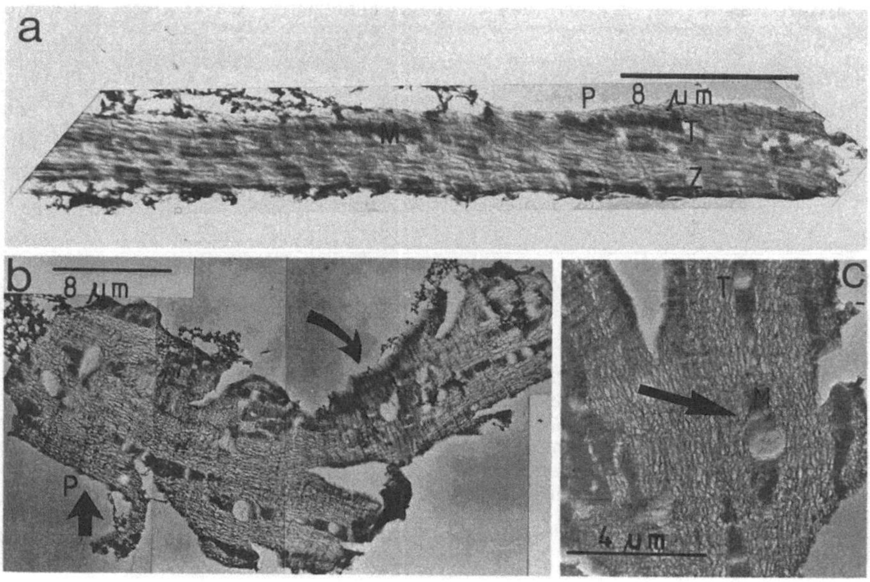

Table 1. Calcium distribution in sarcomeres, jSR, mitochondria and cell border of 2 cells stimulated with 5 pulses a 0.2 s and frozen during the 6th pulse 116 ms after start of depolarisation.

	Cytoplasm	jSR	Cell border	Mitochon.
mean	0.5	0.95	1.96	0.27
SE	0.1	0.1	0.7	0.1
	n=6	n=4	n=9	n=4

The calcium-concentrations in the 4 compartments are simultaneously compared using analysis of variance for unequal sample sizes.The global hypothesis cannot be rejected by $p \geq 0.05$ (By permission from the Am. J. of Physiology). The 95% confidence interval are ±0.3 for cytoplasm, ±0.4 for jSR, ±1.6 for cell border, ±0.1 for mitochondria.

Table 2
Calcium distribution on sarcomerese, jSR, mitochondria and cell border in 3 cells stimulated with 4 pulses a 1 s and shock-frozen 816 ms after start of depolarisation.

	Cytoplasm	jSR	Cell border	Mitochon.
mean	0.35	1.28	2.65	1.25
SE	0.3	0.3	0.7	0.1
	n=13	n=12	n=6	n=12

The analysis of variance for simultaneous control of all contrasts of unequal sizes was carried out according to Scheffe'(1953) and Games (1977). The global hypothesis can be rejected by $p<0.01$. The a posteriori paired comparison between $[Ca]_{Cy}$ and $[Ca]_{SR}$, $[Ca]_{M}$, $[Ca]_{SL}$ showed pair differences at the level of 0.01. $[Ca]_{Cy}$ is not distinguishable from zero (i.e. <100 µmol/kg w.w, which is expected for a completely relaxed normal muscle cell). (By permission from the Am. J. of Physiology). The 95% confidence intervals are ±0.7 for cytoplasm, ±0.6 for jSR, ±1.5 for cell border, ±0.3 for mitochondria.

Fig.6: Freeze-dried cryosections of two single isolated myocytes: a) Longitudinal section of a cell that lays on the pioloform film perfectly parallel to the knife edge of the cryoultramicrotome. Despite the low contrast and the thickness of the section, sarcomeres with Z-lines, (Z), mitochondria (M), and T-tubuli (T) can be identified. During cryocutting the section became deformed.
b) Thinner, internal (3 µm from the cell surface) cryosection of another isolated myocyte, which after freezing "suffered" in some way by the extraction of the electrode. The cell in this case lays on the pioloform film not exactly parallel to the knife edge, therefore the section is not exactly longitudinal. T-tubuli are clearly recognizable and junctional SR adjacent to them is indicated by curved arrow. N: nucleus, P: plasmalemma, M: mitochondria. Arrow: a better frozen region without ice crystal damages, in which the identification of compartments is more difficult. (Note the recognizable surface artifacts due to cutting marks.) (By permission of the Am. J. of Physiol.)

of the entered Ca-ions have moved from the cytosol and have accumulated in the compartments SR, cell border and mitochondria.

Conclusions: 1. We present for the first time a method that can measure Ca-concentrations in compartments of cells in which Ca-influx is defined by voltage-clamp experiments. 2. Ca entry regulates the calcium-load of the cells. 3. Long pulses load cell compartments with higher calcium amounts compared to short pulses.

LITERATURE

Games PA (1977) An improved t table for simultaneous control on g contrasts.J Am Statist Assoc 72:531-534

Fabiato A (1983) Calcium-induced release of calcium from the cardiac sarcoplasmic Am J Physiol 245:C1-C14

Isenberg G and U Klöckner (1982) Calcium currents of isolated bovine ventricular myocytes are fast and of large amplitude. Pflügers Arch 395:30-41

Isenberg G, U Klöckner, D Mascher and U Ravens (1987) Changes in contractility and membrane currents as studied with a single patch electrode whole cell clamp technique. In: Electrophysiology of Single Cardiac Cells. D Noble and T Powell (eds) Academic Press 25-67

Marty A and E Neher (1983) Thight-seal whole-cell recording. In: Sakman and E. Neher (eds) Single channel recording. Plenum Press New York, 107-122,

McDonald TF and W Trautwein (1978) Membrane currents in cat myocardium: separation of inward and outward current. J Physiol 274: 193-210

Moore L, A Schmid and G Isenberg (1984) Linear electric properties of isolated cardiac cells. J Membrane Biol 81:29-40

Reiter M, W Vierlig and K Seibel (1984) Excitation-contraction coupling in rested state contractions of guinea-pig ventricular myocardium. Naunyn-Schmiedeberg's Arch Pharmacol 325:159-169

Scheffe' HR (1953) A method for judging all contrasts in an analysis of variance. Biometrica 40:87-104

Somlyo AV, H Shuman and AP Somlyo (1977) Elemental distribution in striated-muscle and the effects of hypertonicity. Electron probe analysis of cryosections. J Cell Biol 74:828-857

Wendt-Gallitelli MF (1985) Presystolic calcium-loading of the sarcoplasmic reticulum influences time to peak force of contractions. X-ray microanalysis on rapidly frozen guinea-pig ventricular muscle preparations. Basic Res Cardiol 80:617-625

Wendt-Gallitelli MF (1986) Ca pools involved in the regulation of cardiac contraction under positive inotropy. X-ray microanalysis on rapidly frozen ventricular muscles of the guinea-pig. In: Controversal issues in cardiac pathophysiology. Jacob (ed) Basic Res Cardiol 81 Suppl 1: 25-32

Wendt-Gallitelli MF and R Jacob (1984) Effects of non-toxic doses of ouabain on sodium, potassium, calcium distribution in guinea-pig papillary muscle. Electron probe microanalysis. In: Cardiac Glycosides, Receptors and positive Inotropy (Erdman ed) Basic Res Cardiol 79: Suppl 79-87

Wendt-Gallitelli MF, R Jacob and H Wolburg (1982) Intracellular membranes as boundaries for ionic distribution in guinea-pig heart muscle in different defined electromechanical coupling states. Z Naturforsch 37c:712-720

Wheeler-Clark ES, Thormey, JMcD (1987) Electron probe x-ray microanalysis of sarcolemma and junctional sarcoplasmic reticulum in rabbit papillary muscles: low sodium-induced calcium alterations. Circ Res 60: 246-250

Conference Discussion

R. Nassar: Do you have an idea what the sarcomere length is when the myocyte is at rest or clamped at the holding potential before freezing?

M.F. Wendt-Gallitelli: The sarcomere length of the cell can be exactly measured during rest and during the holding potential with a technique described by Gerrit Isenberg in Z. f. Naturforschung, 37c:502-5012, 1982. In rest, the sarcomeres show a length of about 1.85 - 1.88 µm, which is normal for zero preload. During systole, the sarcomeres shorten to at least 1.66 µm, depending on the inotropic state. The isotonic contraction can be recorded till the inverted stage of the microscope is removed from the holder containing the cell immerged in the bath solution. In this way we have information about electrophysiological as well as mechanical parameters of the cell during rest and voltage-clamp. From the moment at which the microscope and the chamber are removed, the electrophysiological information is the only one continuously available till the cell is frozen.

R.A. Steinbrecht: Heptane-embedding the frozen muscle cells perhaps could help in cryosectioning and handling the cryosections. Have you tried?

M.F. Wendt-Gallitelli: Yes, we tried to post-embed the frozen cell in heptane, but the technique did not help, since the frozen cell was surrounded by frozen solution even in this case and it is just the frozen solution which makes cutting very difficult.

B. Gupta: 1. Does your propane cup have a cover and if yes how soon before plunging the tissue is it removed? Ryan from Plymouth has shown that by completely eliminating the cold gas above the cryogen one can get excellent cryopreservation of up to 200 µm. 2. How thick is your silver foil? Titanium foil even a few µm thick is quite rigid for cutting cryosections and reduces both for freezing rate.

M.F. Wendt-Gallitelli: 1. The propane container has a cover which has to be removed ca. 2 s before freezing, since the horizontal movement of the stage has to occur slowly and without vibrations to avoid any movement of the electrode impaling the cell. If the horizontal movement of the stage is faster, the electrode may vibrate and damage the cell. Probably the reproducibility of excellent freezing is limited by the fact that the cell is lying on the pioloform film and, therefore, not in direct contact to the coolant. It is not only the question of optimal freezing in our technique, but cryopreservation must be optimized together with attempts to optimizing records of electrophysiology, timing etc.

2. The thickness of the silver holder (which serves as reference electrode) varies with the distance from the tip to the basis: it is 2-4 μm at the open tip, that it where the cell is lying on the pioloform film. Silver at -140°C is extremely easy to cut, the problem is not to cut the 2 μm-thin silver frame of the holder, but to cut the frozen Tyrode solution in which the cell is bathing during the experiment and which guarantees to the cell a "normal" environment till the moment of freezing.

A.J. Morgan: How much confidence do you have in your ability to measure changes in calcium concentration in the order of 0.5 - 0.7 mM given that the total calcium concentrations in all compartments are close to the detectable limits of the analytical technique.

M.F. Wendt-Gallitelli: The 95 % confidence intervals of the data are given in the legend of Tables 1 and 2. In the case of the data reported in Table 1, due to the small number of measurements, changes in total Ca concentrations in the order of 0.5 - 0.7 mmol/kg d.w. were not significant at the 0.05 level. But, as shown in Table 2, if in a compartment 0.5 mmol/kg d.w. were measured and in the adjacent compartment 2.65 mmol/kg d.w. and in a third compartment 1.8 mmol/kg d.w. were detected, and analysis of variance for unequal contrast reveal significant differences at least at the 0.05 level between these concentrations, the differences can be considered biologically consistent. Concerning the statistical error associated with the single Ca measurement, if a single measurement of 1 mmol/kg d.w. has a statistical error of 1 mmol/kg d.w. or more, the measured concentration is zero. If, however, a single measurement of 1 mmol/kg d.w. has, for instance, a statistical error of 0.5 or less, my confidence in the ability to measuring changes in calcium concentration in this range increases as the statistical error associated with the single measuremnt decreases.

X-Ray Microanalysis of Fast Exocytotic Processes

Karl Zierold, Imme Gerke[1)], Manfred Schmitz[2)]
Max-Planck-Institut für Systemphysiologie
Rheinlanddamm 201
D-4600 Dortmund 1
F.R.G.

INTRODUCTION

Fast freezing of biological cells, called cryofixation, arrests physiological processes and makes them accessible to investigation by electron microscopy and electron probe microanalysis. We have used time-controlled cryofixation to study morphological alterations and elemental shifts during fast exocytosis of organelles. Two examples: Exocytosis of trichocysts in <u>Paramecium caudatum</u> and discharge of nematocysts in <u>Hydra attenuata</u> are considered in more detail. Although the mechanisms of both processes are not yet understood, X-ray microanalysis of cryosections from cells frozen after defined time intervals following stimulation provides data on element distribution, which might help to elucidate the underlying biological functions. Finally, possibilities and limitations of time-controlled cryofixation are discussed with respect to diffusion length of ions and freezing time.

---

1) Zoologisches Institut, Universität Zürich, Zürich, Switzerland
2) Bundesministerium für Jugend, Familie, Frauen und Gesundheit, Bonn, F.R.G.

MATERIALS AND METHODS

Cryofixation was done by plunging the specimens prepared on modified Balzers gold planchettes into liquid propane cooled by liquid nitrogen.

Paramecia were collected from their nutrition medium by a pipette and placed within a droplet of less than 0.5 mm in diameter upon the gold planchette. This specimen was shot by a crossbow-like device into liquid propane with a velocity of 10 m/s, monitored by photosensors. Trichocysts were stimulated for exocytosis by shooting the specimen through a liquid film of 0.015 % AED (aminoethyldextran), spanned by a wire loop above the cooling agent (Fig. 1a). AED was reported to stimulate exocytosis of trichocysts in paramecia (Plattner et al., 1984, 1985). Time intervals between stimulation and cryofixation were defined by adjusting the distance between the liquid film and the cooling liquid.

One hydra animal was placed on a gold planchette in such a way that tentacles were exposed to the top. This specimen was shot into liquid propane by an air pressure device resulting in a specimen velocity of 1 m/s. 10 mm above the cooling agent a flexible copper foil was mounted with an electrical potential of 20 V against the specimen holder (Fig. 1b). Before reaching the cooling agent, the tissue touched the copper foil resulting in an electrical current which stimulated discharge of nematocysts in the tissue.

After cryofixation, 100 nm thick cryosections were prepared by a dry glass knife at a temperature below 170 K. The sections were cryotransferred into the cold stage of a STEM (scanning transmission electron microscope). Freeze-drying was done in the cryotransfer chamber at a temperature of approximately 190 K. X-ray microanalysis was done by an energy dispersive spectrometer combined with a multi-channel analyzer. Data were evaluated according to the peak-to-background method

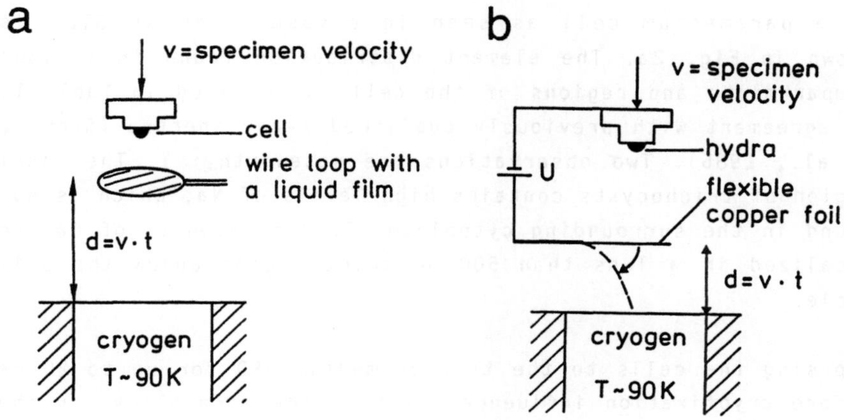

Fig. 1
Time-controlled cryofixation to study fast exocytotic processes.
a) Paramecia in droplets on gold planchettes are shot into the cryogen through a liquid film containing the trigger agent AED. Time interval (t) between stimulation and cryofixation is defined by adjusting the distance d (from Schmitz and Zierold, 1989).
b) Tentacles of hydra mounted on gold planchettes touch a flexible copper foil before plunging into the cryogen. Between the specimen and the copper foil the voltage U = 20 V is applied.

(Hall and Gupta, 1982). Details of the preparation and analysis technique have been described by Zierold (1986, 1988).

TRICHOCYST EXOCYTOSIS IN PARAMECIUM CAUDATUM

Paramecia are ciliates living in aqueous medium. They contain specialized elliptic organelles called trichocysts, which are expanded to long rods and ejected like arrows from the cell after stimulation. We have studied morphological changes and ion shifts related to this exocytotic process. The morphology of a paramecium cell as seen in cryosections in STEM is shown in Fig. 2a. The element distribution found in various compartments and regions of the cell is compiled in Table 1, in agreement with previously published X-ray spectra (Schmitz et al., 1985). Two observations are noteworthy: 1. The basal region of trichocysts contains high levels of Na, which is not found in the surrounding cytoplasm. 2. High amounts of Ca are localized in a less than 500 nm thick region below the pellicle.

Exposing the cells to the trigger medium AED for up to 20 ms before cryofixation influences neither the morphology nor the intracellular element distribution. Fig. 2b shows a trichocyst docked at the cell membrane 20 ms after exposure of the cell to AED. 30 ms after contact of the cell with AED, trichocysts are expanded to long rods like arrows and expelled from the cell, at least at the cellular surface exposed to the liquid film containing AED (Fig. 2c). The remaining sacs called trichocyst ghosts exhibit high amounts of Mg and S, whereas Na has left the cell. Further experiments have revealed Ca in trichocyst sacs during exocytosis, whereas extracellularly offered divalent ions such as Ba and Sr did not enter the cell (Schmitz and Zierold, 1989).

Our results indicate that trichocysts in paramecia might be involved in cellular osmoregulation. Ca ions claimed to be necessary for trichocyst decondensation (Garofalo et al., 1983) could be provided by Ca stores found below the pellicle.

→

Fig. 2 (opposite page)
Freeze-dried cryosections from Paramecium caudatum:
a) Resting state, b) 20 ms after stimulation by AED, c) 30 ms after stimulation by AED.
C = cytoplasm, E = extracellular space, N = nucleus, P = pellicle, T = trichocyst, T_b = basal region of trichocyst, T_g = trichocyst ghost, T_t = trichocyst tip.

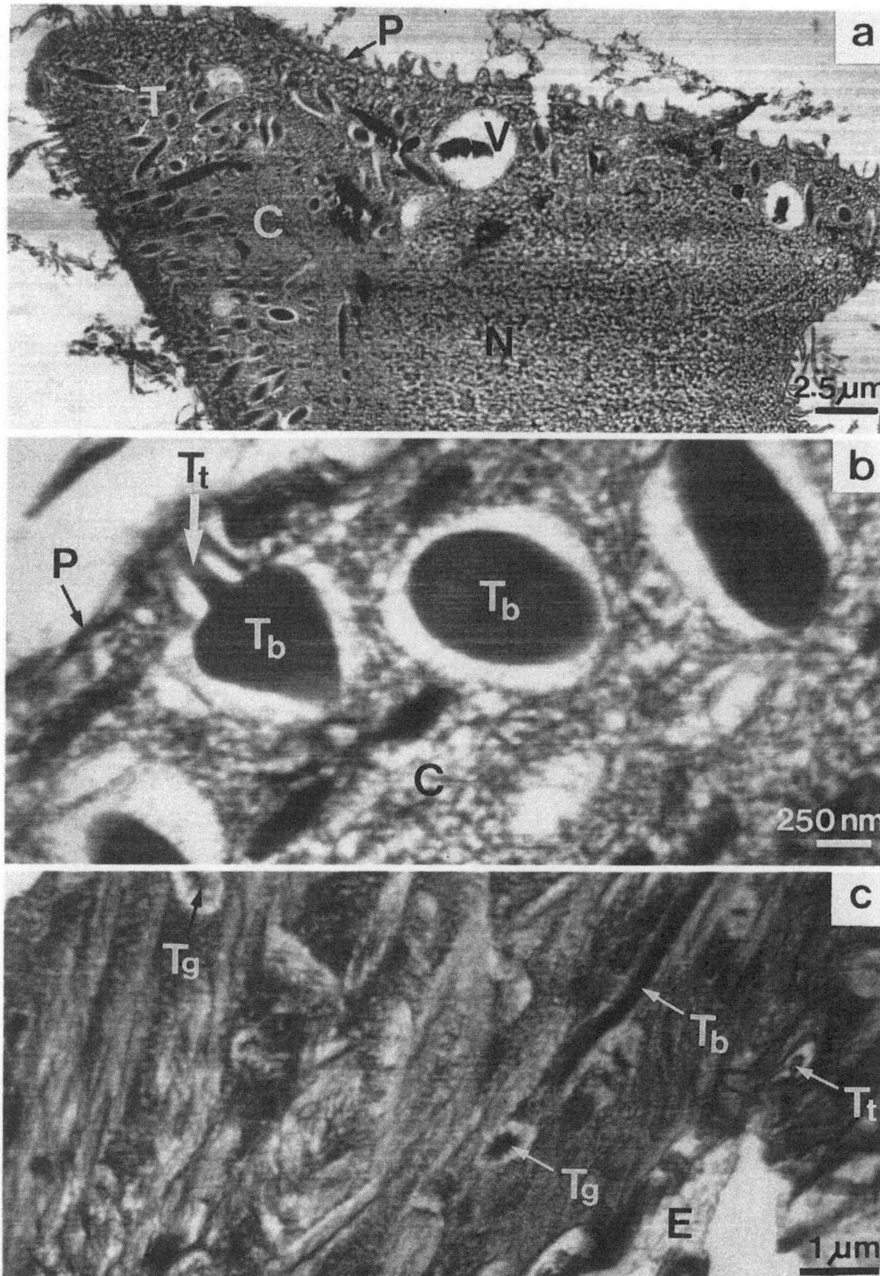
a
P
T
V
C
N
2.5 μm
b
Tt
P
Tb
Tb
C
250 nm
c
Tg
Tb
Tt
Tg
E
1 μm

Table 1

Element distribution in various compartments of *Paramecium caudatum* cell, in mmol/kg dry weight ± standard deviation, n = number of measurements. Horizontal bars indicate concentrations below the detection limit. From: Schmitz and Zierold (1989)

compartment	n	Na	Mg	P	S	Cl	K	Ca
cytoplasm	15	---	104±96	489±225	163±82	30±20	125±70	---
cytoplasm, adjacent to pellicle	10	56±26	77±43	376±192	169±65	72±110	82±47	267±238
trichocyst base	35	177±59	21±26	24±47	81±31	20±32	31±42	---
trichocyst tip	8	41±36	---	319±235	200±102	---	63±68	---
trichocyst ghost	29	---	821±585	---	1469±631	53±70	55±80	---

NEMATOCYST DISCHARGE IN HYDRA ATTENUATA

Nematocysts of the stenotele type are stinging capsules which can be discharged from specialized cells called nematocytes in hydra within 3 ms after stimulation. However, the stylets of those capsules are ejected within 50 μs, whereas the adhering tubule follows within milliseconds (Holstein and Tardent, 1984). The morphology of this process was described by Tardent and Holstein (1982). Fortunately, not all nematocysts are discharged immediately after stimulation, but with varying time delay in the range of several milliseconds. This delay of the individual secretion processes results in cryofixation of different states during nematocyst discharge. In the resting state cryosectioned nematocysts appear electron dense and poor in structural details (Fig. 3a). After discharge upon electrical stimulation, the matrix of nematocysts looks less electron dense than in the resting state, and in some cases parts of the tubule can be identified (Fig. 3b). The nematocysts in the cryosection shown in Fig. 3c exhibit different states during discharge. Element concentrations measured in cryosections from nematocytes frozen in the resting state and 10 ms after electrical stimulation are compiled in Table 2. Comparison of the data reveals two striking results: 1. The high K content of the nematocysts in the range of 2 mol/kg dry weight in the resting and stimulated state. 2. After stimulation, Ca increases in the wall and in the matrix of nematocysts, whereas S is decreased.

The data in Table 2 are different from those reported by Lubbock et al. (1981). These authors found by X-ray microanalysis of 1-3 μm thick cryosections a considerable loss of Ca in nematocysts of sea anemones during discharge. At present, we cannot explain our data with respect to the mechanism of nematocyst discharge. The K accumulation in the matrix of nematocysts could be related to high osmotic pressure supposed to be responsible for nematocyst discharge triggered by Ca influx due to electrical stimulation.

CRYOFIXATION AND ION SHIFTS

Cryofixation permits the capture of fast processes in cells. However, diffusible ions may undergo redistribution immedately before cryofixation. The mean diffusion length, l, can be roughly estimated by the equation

$$l = \sqrt{2 \cdot D \cdot t} \qquad (1)$$

with D = diffusion constant and t = diffusion time.
The used techniques shown in Fig. 1 indicate the shortest time interval achievable between stimulation and freezing to be about 1 ms. The diffusion constant of highly mobile ions such as Na, K, Cl in water at 300 K is approximately $2 \cdot 10^{-9}$ m^2/s. According to equation (1) the mean diffusion length is 2 µm within 1 ms, 6 µm within 10 ms, and 10 µm within 30 ms. As the jump frequency of water molecules drops from 10^{11} per second to 10^6 per second by freezing, diffusion can be neglected below the freezing point. Although ion diffusion is probably reduced slightly in presence of macromolecules and solutes, the diffusion length of diffusible ions in aqueous solutions has to be taken into account, in particular in cells exhibiting fast functional changes. Localization of ion gradients with a spatial resolution of less than 1 µm seems to be unreliable, unless ion mobility is reduced by chemical or electrostatic interaction with macromolecules.

ACKNOWLEDGEMENT

We thank Mrs. S. Dongard for her careful technical assistance.

Fig. 3 (opposite page)
Freeze-dried cryosections from Hydra attenuata.
a) Resting state, b) and c) cryofixed 10 ms after electrical stimulation. C = cytoplasm, N = nematocyst, T = tubule, W = nematocyst wall.

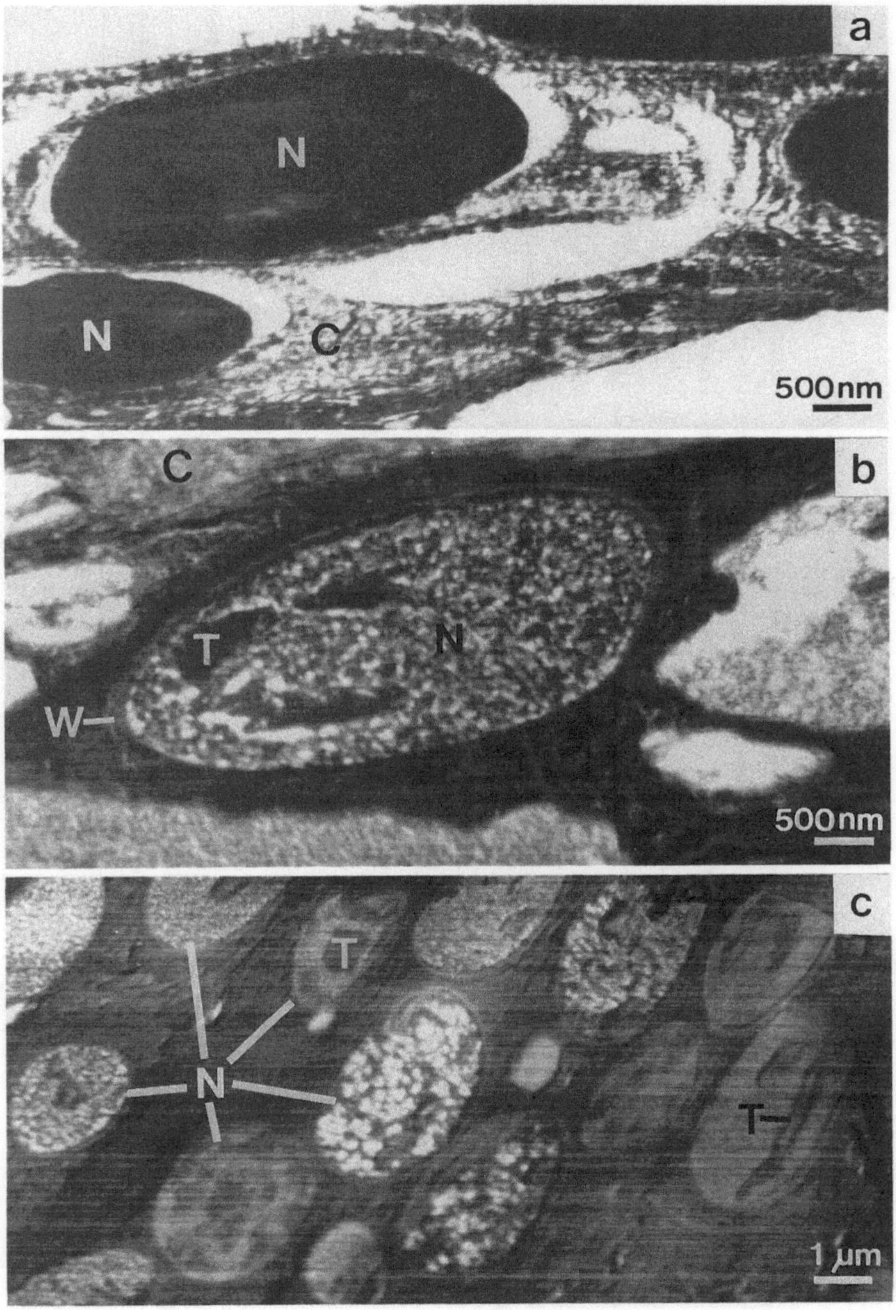
a
N
N
C
500 nm
b
C
T
N
W
500 nm
c
T
N
T
1 µm

Table 2

Element concentrations (mol/kg dry weight ± standard deviation) measured in nematocytes of _Hydra attenuata_ cryofixed in the resting state and 10 ms after electrical stimulation. n = number of measurements. Significance levels found in the stimulated state in comparison to the resting state are denoted by: * = 0.05, ** = 0.01, *** = 0.001.

		n	Na	Mg	P	S	Cl	K	Ca
Resting state:	cytoplasm	18	0±1	51±22	406±113	231±65	39±24	127±65	21±21
	nematocyst wall	15	6±10	14±14	72±91	452±213	12±10	362±458	1±2
	nematocyst	19	60±42	22±16	11±15	123±67	13±8	2203±905	6±9
	tubule	7	47±32	16±14	21±44	342±131	3±4	627±325	4±4
Stimu-lated state:	cytoplasm	11	9±15*	43±16	302±104*	173±39*	9±3***	92±59	23±17
	nematocyst wall	8	56±73*	56±66*	97±68	336±156	7±3	420±314	23±21***
	nematocyst	18	125±100*	53±74	3*±5	45±30***	9±6	2571±483	20±15**

REFERENCES

Garofalo RS, Gilligan DM, Satir BH (1983) Calmodulin antagonists inhibit secretion in Paramecium. J Cell Biol 96:1072-1081

Hall TA, Gupta BL (1982) Quantification for the X-ray microanalysis of cryosections. J Microsc 126:333-345

Holstein T, Tardent P (1984) An ultra high-speed analysis of exocytosis: nematocyst discharge. Science 223:830-833

Lubbock R, Gupta BL, Hall TA (1981) Novel role of calcium in exocytosis: Mechanism of nematocyst discharge as shown by X-ray microanalysis.

Plattner H, Matt H, Kersken H, Haacke B, Stürzl R (1984) Synchronous exocytosis in Paramecium cells. 1. A novel approach. Exp Cell Res 151:6-13

Plattner H, Stürzl, R, Matt H (1985) Synchronous exocytosis in Paramecium cells. IV. Polyamino compounds as potent trigger agents for repeatable trigger-redocking cycles. Eur J Cell Biol 36:32-37

Schmitz M, Meyer R, Zierold K (1985) X-ray microanalysis in cryosections of natively frozen Paramecium caudatum with regard to ion distribution in ciliates. Scan Electron Microsc 1985/I:433-445

Schmitz M, Zierold K (1989) X-ray microanalysis of ion changes during fast processes in cells, as exemplified by trichocyst exocytosis of Paramecium caudatum. In: Plattner H (ed.) Electron Microscopy of Subcellular Dynamics. CRC Press, Boca Raton Fl USA, in press

Tardent P, Holstein Th (1982) Morphology and morphodynamics of the stenotele nematocyst of Hydra attenuata Pall. (Hydrozoa, Cnidaria). Cell Tiss Res 224:269-290

Zierold K (1986) Preparation of cryosections for biological microanalysis. In: Müller M, Becker RP, Boyde A, Wolosewick JJ (eds.). The Science of Biological Specimen Preparation 1985, Scanning Electron Microsc Inc, AMF O'Hare, Ill, pp 119-127

Zierold K (1988) X-ray microanalysis of freeze-dried and frozen-hydrated cryosections. J Electron Microsc Techn 9:65-82

Conference Discussion

H.K. Hagler: In the region (500 nm) below the pellicle were you able to observe any localized redistribution of Ca which correlates with the discharge of nearby trichocysts?
K. Zierold: No, we found similar amounts of Ca below the pellicle in stimulated and in unstimulated cells. We have also not yet succeeded in identifying specific Ca accumulating structures in cryosections of paramecium.

R.A. Steinbrecht: Where does all the magnesium come from which you find in the empty trichocyst ghosts of paramecium after stimulation? There are no larger magnesium reservoirs in the unstimulated cells.
K. Zierold: The origin of the high Mg and S concentrations found in trichocyst ghosts is not clear. However, it should be taken into account that these amounts were found in small precipitates within the ghost. The mean concentrations averaged over the whole ghost area would be much lower than the values compiled in Table 1.

B.L. Gupta: (1) In estimating rates of diffusion in water one must not forget the viscosity factor. In unstirred water layer rates of diffusion can be 10 to 1000 times slower. (2) May I remind that in isolated nematocysts of hydra one can simulate many of the fast events of discharge. However, these nematocysts in vitro have no K at all, but instead have up to 2 M kg^{-1} dry wt of Mg.
K. Zierold: (1) The given estimate on ion diffusion length describes the "worst" case: Diffusion in an aqueous compartment without any hindrance by other molecules or surfaces. In real biological systems ion diffusion may be lower than estimated from the diffusion constants in pure water. (2) We found that isolated nematocysts are able to store a wide variety of cations (also Mg) depending on the composition of the surrounding medium.

Electron Probe Microanalysis in Pathology

Godfried M. Roomans
Department of Human Anatomy
University of Uppsala, Box 571
S-75123 Uppsala, Sweden

The early literature on the use of X-ray microanalysis in pathology has been reviewed by Roomans (1983) and more recently by Baker et al. (1985) and Roomans (1988). From these reviews it appears that the analysis and identification of exogenous particles, especially in lung, is quantitatively the most important application. Apart from the clinical importance of this field, the relative ease of preparation and analysis of samples for such studies is responsible for this situation. A second group of applications concerns the analysis of pathological local accumulation of metals (copper, iron) in cells. A third group concerns pathological deposits of calcium in cells and tissues, and a fourth group disease-related redistribution of diffusible ions.

Preparation of samples for X-ray microanalysis in pathology

For analysis of particulates in lungs, the tissue samples are often digested with acid, and the particulates (e.g., asbestos fibers) are collected onto a film-covered grid or a filter. If the localization of the fiber in the tissue is relevant, deparaffinized sections of fixed and paraffin-embedded material are used (Abraham and Burnett 1983).

Generally, analysis of pathological accumulations of e.g., iron or copper is carried out on conventionally fixed and embedded tissue. The possibility of elemental losses during preparation has rarely been considered. Even studies of pathological calcifications are often carried out on fixed material. In many cases, the microanalysis may have been planned only after the tissue had already been collected, fixed and embedded for conventional light or electron microscopy. In addition, few of these studies were quantitative, so that elemental loss was not relevant as long as the element of interest still was present in a detectable amount.

Studies concerning disease-related redistribution of diffusible ions are relatively rare. For such studies, cryopreparation techniques have to be used and collection and preparation of the samples has to be carefully planned. It is well known that freezing is the most critical step of the preparative procedure and that even under optimal conditions only an outer layer of about 30 μm can be frozen without appreciable ice crystal damage. Such optimal conditions for freezing can be obtained in work with experimental animals under laboratory conditions, but generally not in situations where human tissue has to be removed. According to our experience, the following complications can occur:

(1) Sampling and freezing may have to be done in connection with surgery. Restrictions may apply with regard to the technique of freezing and delays between the actual removal of the tissue and the freezing may be unavoidable.

(2) The pathological changes may not be evenly distributed within the tissue; the well-frozen outer layer may be irrelevant to the purpose of the study.

(3) Local or general anesthesia may affect the ion distribution; the microanalyst will not have control over the type of anesthetic used and the effect of anesthetics on ion distribution has not been well studied.

The need to take tissue samples as soon as possible after death in the case of autopsy material has been illustrated in a study by Roomans and Wroblewski (1985). Often, freezing will have to be performed under suboptimal conditions and fairly extensive ice crystal damage may therefore have to be accepted. X-ray microanalysis at the cellular level using 2-16 μm thick cryostat sections may still be possible (Wroblewski et al. 1988). Sections of freeze-substituted or freeze-dried embedded samples may provide good alternatives to ultrathin cryosections: even relatively poorly frozen material can be sectioned, and tissue blocks can be more easily reoriented if the structural inhomogeneity of the block makes this necessary (Wroblewski et al. 1988).

Microanalysis of exogenous particles

A number of diseases has been associated with inhalation of asbestos fibers: asbestosis (pulmonary interstitial fibrosis), diffuse (malignant) mesothelioma of pleura and peritoneum, parietal pleural plaques, and bronchiogenic carcinoma. Other diseases related to the (industrial) environment include silicosis, and talc, coal, and mixed dust pneumoconiosis. X-ray microanalysis can aid in the diagnosis of the disease and can be particularly important from a medico-legal viewpoint (Abraham 1980). In addition, microanalysis is used in epidemiological studies on particulate burden in lung tissue, comparing populations from different environments.

Pathological accumulation of exogenous and endogenous metals

X-ray microanalysis has demonstrated pathological accumulation of exogenous or endogenous metals mainly in the lysosomes. If toxic metals remain dissolved in the cytoplasm to an appreciable extent, their effect is fatal long before a concentration detectable by X-ray microanalysis is reached. The presence of thorium in various cells in cancer patients is an example of an iatrogenically induced metal accumulation. Thorotrast, a thorium dioxide containing radio-opaque substance was previously used in radiology but causes tumors after a variable latency period. Iron and copper are the two elements most commonly implicated in pathological metal accumulations of endogenous origin; X-ray microanalysis shows these metals to be mainly localized in lysosomes. Iron deposits are very common in a variety of tissues after bleeding (hemosiderosis) and numerous case reports can be found in the literature. Abnormal accumulation of copper in various organs occurs in Wilson's disease and one of the first applications of X-ray microanalysis to pathology actually concerned this disease (Tousimis and Adler 1963).

Calcium deposits

X-ray microanalysis can be used to analyze urinary stones; knowledge of the composition of the stones may be important in deciding on a treatment to minimize stone recurrence. Kim et al. (1986) showed that X-ray microanalysis was very cost effective and more reliable than other analytical methods commonly used for stone analysis.

Accumulation of calcium in tissue associated with a pathological process is a general phenomenon. Calcification is particularly common in "ischemia-reflux" situations where despite tissue damage, circulation remains intact or is restored. Atherosclerosis is quantitatively the most important disease related to pathological calcifications. At the cellular level, however, the process leading to calcification is still incompletely understood.

Disease-linked redistribution of diffusible ions

Redistribution of diffusible ions is found associated with a number of diseases. In some cases, the redistribution is not specific but symptomatic of generalized cell damage. In other cases, abnormal ion distribution is closely linked to the primary defect underlying the disease.

Cystic fibrosis: Cystic fibrosis (CF) is the most common lethal, hereditary disease among the Caucasian population. The main clinical symptoms are chronic respiratory disease associated with viscous mucus blocking the airways, and pancreatic insufficiency. Elevated levels of Na and Cl in the sweat are a diagnostic sign of the disease. At present it is widely believed that a defect in transepithelial chloride and water transport is closely related to the basic defect, but also an abnormal regulation of intracellular calcium levels has been implicated.

X-ray microanalysis can be used in the diagnosis of CF. Diagnosis is carried out by chemical analysis of pilocarpine-induced sweat; in some countries, however, neutron activation analysis of nail clippings (on which sweat has collected) is used. X-ray microanalysis of nail clippings is an alternative to this latter method; Na and Cl levels in clippings from CF patients are markedly higher than and can easily be distinguished from those of healthy subjects (Roomans et al. 1978).

Elevated calcium levels have been found in many exocrine glands and also in other cells of CF patients. X-ray microanalysis showed elevated calcium levels in cultured fibroblasts (Roomans 1986), confirming results obtained by other methods, but the relevance of these findings is still not clear. X-ray microanalysis of sections of freeze-substituted bronchial biopsies (obtained during bronchoscopy) showed higher levels of calcium in the goblet cells of CF patients compared to a control group of patients with chronic bronchitis (Roomans et al. 1986).

Recently we have carried out analysis on cultured respiratory epithelial cells from CF patients. Under resting conditions, the ion content of the cells is not significantly different from that of cultured cells from control donors. However, after ß-adrenergic stimulation, a difference between CF cells and control cells becomes apparent (Fig. 1a). In CF cells, apparently chloride uptake over the basal membrane (via a $Na^+/K^+/2Cl^-$ cotransport system) is functional, but not the cAMP-regulated chloride efflux over the apical membrane (Roomans et al. 1988). This agrees well with electrophysiological studies indicating that in CF cells the cAMP activation of the chloride channel may be defective. Because of indications that in CF cells Na^+ uptake over the apical membrane might be more sensitive to amiloride than in control cells, we also tested the effect of amiloride on the elemental composition of the cells (Fig. 1b), and found indeed a

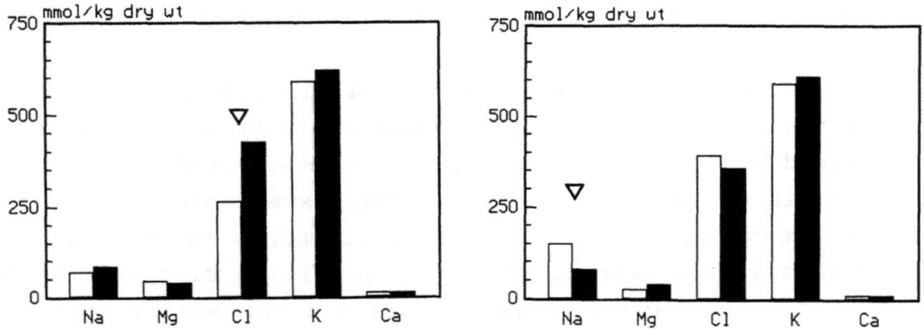

Fig. 1: X-ray microanalysis of cultured respiratory epithelial cells from CF patients (solid bars) and controls (open bars). Significant differences are indicated by triangles. The cultures were initiated from polyps, and the cells were cultured on membrane filters and mounted in an Ussing chamber to allow electrophysiological studies. In (a) 8-Br-cAMP (5 mM), in (b) amiloride (50 μM) were added to the cells. After 3 min, the cell cultures were washed with isotonic mannitol to remove the extracellular medium, frozen in liquid nitrogen, freeze-dried and analyzed in a scanning electron microscope at 20 kV (Roomans et al. 1988).

marked effect of amiloride on CF cells but not on control cells (Roomans et al. 1988).

Based upon an earlier theory linking CF to a defect in the nervous regulation of exocrine gland function, Martinez proposed the chronically reserpinized rat as an animal model for CF, since the salivary glands and to a certain extent also other tissues showed some functional abnormalities resembling those found in CF patients (reviewed in Martinez and Barbero 1985). In a series of studies we carried out X-ray microanalysis on exocrine glands of the reserpinized rat and other animal models based on interference with nervous regulation of gland function (Roomans et al. 1982, Müller and Roomans 1985). A consistent finding was the abnormally high level of calcium in the submandibular acinar cells (Fig. 2a), partly caused by an increase in the local calcium concentration of the mucus, and partly by an increase in the relative mucus content of the cells. However, different types of interference with either the nervous regulation or the

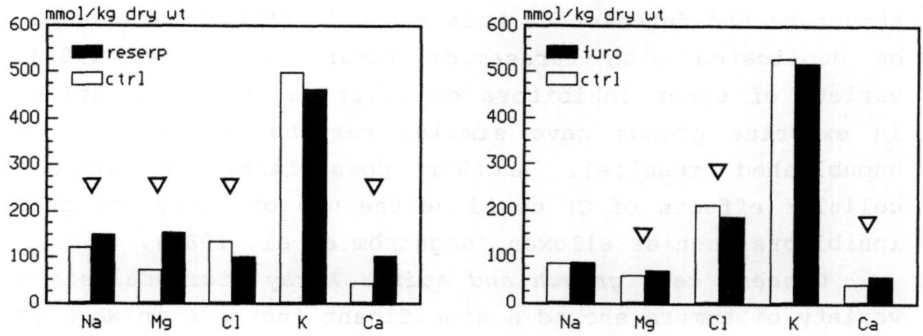

Fig. 2: X-ray microanalysis of rat submandibular gland acinar cells. In (a) female reserpinized rats (solid bars) are compared to female control rats (Roomans et al. 1982, Müller and Roomans 1985), in (b) male rats chronically (1 month) treated with furosemide are compared to control rats kept for the same period on standard food (Scarlett et al. 1988).

of interference with either the nervous regulation or the metabolism of the gland caused similar effects on calcium levels and no firm conclusions as to the relevance of the increased calcium could be drawn (Roomans 1986).

If a defective transepithelial Cl^- and water transport is a central factor in the ethiology of CF, it should theoretically be possible to evoke CF-like symptoms in experimental animals by chronically inhibiting transepithelial Cl^- transport. Chronic (1 month) treatment of rats with furosemide caused a significant inhibition of fluid transport in the submandibular gland. X-ray microanalysis showed a significant decrease of the cellular chloride concentration (in accordance with the inhibition of chloride uptake caused by furosemide) and an increase in the cellular calcium concentration due to mucus accumulation (Scarlett et al. 1988). The pattern of elemental changes (Fig. 2b) is actually rather similar to that found in chronically reserpinized rats in which also a defective chloride and water transport could be demonstrated (Martinez and Barbero 1985). Significant but relatively minor changes could also be found in the pancreas of furosemide-treated animals but the severe changes in structure and function of this gland in CF patients could not be duplicated with furosemide (Scarlett et al. 1988). A variety of other inhibitors of chloride and water transport in exocrine glands gave similar results (Sagström et al., unpublished results). Another possibility to mimic the cellular effects of CF could be the use of adenylate cyclase inhibitors such as alloxan (Sagström et al. 1987).

Cancer, cell growth and aging: X-ray microanalysis on a variety of tumors showed a significant increase in Na/K ratio (Zs. Nagy et al. 1981, Wroblewski et al. 1983). Cameron and Smith (1980) showed increased levels of Na, Mg, P, and K in a variety of rapidly dividing, non-neoplastic cells, when compared to similar, but slowly dividing cells. Grundin et al. (1985) found increased levels of Mg, P, and K in the epidermal cells (stratum germinativum and stratum spinosum) of patients with psoriasis. A sustained increase in Na/K ratio thus appears to be specific for neoplasia but not for cell proliferation in general. In aging cells, the concentrations of all electrolytes appears to increase; according to some, the water permeability of the plasma membrane is changed, leading to increased electrolyte concentrations in cytoplasm and nucleus (Lustyik and Zs. Nagy 1985), according to others, changes in the water permeability of the mitochondrial membrane are more relevant to the aging process (von Zglinicki 1987).

Skeletal muscle disorders: In a series of studies, Wróblewski and Edström (1984) investigated the elemental composition of muscle fibers in a number of skeletal muscle disorders. Different disorders resulted in different changes of the elemental concentrations, even if not all patterns of change were specific for a particular disorder.

Zinc in spermatozoa: X-ray microanalysis has shown a high content of zinc in the sperm nucleus, which could contribute to the stability of the quaternary structure of the nucleoproteins (Björndahl et al. 1986). Abnormally low concentrations of zinc in sperm nuclei could explain some cases of infertility in men living in barren unions where no other abnormality was apparent (Kvist et al. 1987).

This work was supported in part by grants from the Swedish Medical Research Council, the Swedish National Association against Heart and Lung Diseases, and the Swedish Association for Cystic Fibrosis.

References

Abraham JL (1980) Biomedical microanalysis -- Putting it to work now in diagnostic pathology. Scanning Electron Microsc 1980/IV: 171-178.

Abraham JL, Burnett BR (1983) Quantitative analysis of inorganic particulate burden *in situ* in tissue sections. Scanning Electron Microsc 1983/II: 681-696.

Baker D, Kupke KG, Ingram P, Roggli VL, Shelburne JD (1985) Microprobe analysis in human pathology. Scanning Electron Microsc 1985/II: 659-680.

Björndahl L, Kjellberg S, Roomans GM, Kvist U (1986) The human sperm takes up zinc at ejaculation. Int J Androl 9: 77-80.

Cameron IL, Smith NKR (1980) Energy dispersive X-ray microanalysis of the concentration of elements in relation to cell reproduction in normal and in cancer cells. Scanning Electron Microsc 1980/II: 463-474.

Grundin TG, Roomans GM, Forslind B, Lindberg M, Werner Y (1985) X-ray microanalysis of psoriatic skin. J Invest Dermatol 85: 378-380.

Kim KM, Alpaugh HB, Johnson FB (1985) X-ray microanalysis of urinary stones, a comparison with other methods. Scanning Electron Microsc 1985/III: 1239-1246.

Kvist U, Björndahl L, Kjellberg S (1987) Sperm nuclear zinc, chromatin stability, and male fertility. Scanning Microsc 1: 1241-1247.

Lustyik G, Zs.Nagy I (1985) Alterations of the intracellular water and ion concentration in brain and liver cells during aging as revealed by energy dispersive X-ray microanalysis of bulk specimens. Scanning Electron Microsc 1985/I: 323-337.

Martinez JR, Barbero GJ (eds) (1985) Animal models for cystic fibrosis: the reserpine treated rat. San Francisco Press, San Francisco.

Müller RM, Roomans GM (1985) X-ray microanalysis of exocrine glands in animal models for cystic fibrosis. Scanning Electron Microsc 1985/IV: 1583-1601.

Roomans GM (1983) Energy-dispersive X-ray microanalysis. In: Electron Microscopy in Human Medicine (Johannessen JV, ed), McGraw-Hill, London, vol 11B, pp 89-156.

Roomans GM (1986) Calcium and cystic fibrosis. Scanning Electron Microsc 1986/I: 165-178.
Roomans GM (1988) X-ray microanalysis in clinical research. In: Laschi R (ed) Electron microscopy in medicine. Editrice Compositori, Bologna, in press.
Roomans GM, Wroblewski J (1985) Post-mortem storage of tissue for X-ray microanalysis in pathology. Scanning Electron Microsc 1985/II: 681-686.
Roomans GM, Afzelius BA, Kollberg H, Forslind B (1978) Electrolytes in nails analyzed by X-ray microanalysis in electron microscopy. Considerations of a new method for the diagnosis of cystic fibrosis. Acta Paediatr Scand 67: 89-94.
Roomans GM, Wei X, Ceder O, Kollberg H (1982) The reserpinized rat in the study of cystic fibrosis: X-ray microanalysis of submandibular gland and pancreas. Ultrastruct Pathol 3: 285-293.
Roomans GM, von Euler AM, Müller RM, Gilljam H (1986) X-ray microanalysis of goblet cells in bronchial epithelium of cystic fibrosis patients. J Submicrosc Cytol 18: 613-615.
Roomans GM, Sagström S, Wróblewski R, Keulemans JLM, Bijman J (1988) X-ray microanalysis of respiratory epithelial cells in cystic fibrosis. Proc Annual SCANDEM Meeting, Aarhus, J Ultrastruct Mol Struct Res, in press.
Sagström S, Scarlett SM, Sagulin GB, Roomans GM (1987) Early effects of alloxan on rat submandibular gland. J Submicrosc Cytol 19: 555-559.
Scarlett SM, Sagström S, Sagulin GB, Roomans GM (1988) Effects of chronic furosemide treatment on rat exocrine glands. Exp Mol Pathol 48: 206-215.
Tousimis AJ, Adler I (1963) Electron probe X-ray microanalyzer study of copper within Descemet's membrane of Wilson's disease. J Histochem Cytochem 2: 40-47.
Von Zglinicki T (1987) A mitochondrial membrane hypothesis of aging. J Theor Biol 127: 127-132.
Wroblewski J, Anniko M, Wróblewski R (1983) Healthy and diseased epithelium of the oral and maxillary sinus mucosa studied by X-ray microanalysis. J Submicrosc Cytol 15:593-601.
Wroblewski J, Wróblewski R, Roomans GM (1988) Low temperature techniques for X-ray microanalysis in pathology: Alternatives to cryoultramicrotomy. J Electron Microsc Tech 9: 83-98.
Wróblewski R, Edström L (1984) Recent applications of X-ray microanalysis in muscle pathology. Scanning Electron Microsc 1984/I: 249-259.
Zs.Nagy I, Lustig G, Zs.Nagy V, Zarándi B, Bertoni-Freddari C (1981) Intracellular Na^+:K^+ ratios in human cancer cells as revealed by energy-dispersive X-ray microanalysis. J Cell Biol 90: 769-777.

Conference Discussion

R. Rick: Patch-clamp studies in airway epithelia seem to suggest that it is the cAMP-dependent regulation of the Cl channel which is defective in cystic fibrosis. Do you have data that support his notion?

G.M. Roomans: Our data agree with the notion that the defective transepithelial chloride transport in cystic fibrosis patients is due to a defect in the apical, rather than in the basal membrane. The data indicate that the chloride channel in the cystic fibrosis cells cannot be activated by cAMP. However, the experiment does not distinguish between the possibility of on one hand a defective channel, or on the other hand a defective regulation of the channel.

R.A. Steinbrecht: Have you also tried element ratios instead of absolute concentrations in those cases where variation of data was considerable, and did this reduce the variation? In other words, are there still problems with background correction in your analysis?

G.M. Roomans: In analysis of thin specimens, the extraneous contribution to the background may give problems in quantitation, probably because the simple model for extraneous background generation used in the correction normally used is not adequate. In those cases, determination of elemental ratios may result in a smaller standard deviation, which may make it easier to distinguish e.g., between a "control" and a "diseased" group. As an example, in the analysis of zinc in spermatozoa, we prefer to use the Zn/S ratio instead of the absolute zinc concentration for this reason. However, in the quantitative analysis of thick specimens this problem is of much less importance, because the extraneous contribution is neglibible. Here, in effect, X-ray absorption may make the use of peak-to-background ratios more accurate than the use of elemental ratios based on characteristic counts only.

H.K. Hagler: The presentation of data without indicating how the data was sampled, the number of samples, the basic statistics and how significance was measured does not contribute data which can be used by other investigators. Would you outline your sampling and give more complete statistical descriptions of your data?

G.M. Roomans: The details requested are out of scope in a review such as this; moreover, they have been presented in full in the original papers (e.g., Roomans et al. 1982, Müller and Roomans 1985, Grundin et al. 1985, Roomans et al. 1986, Kvist et al. 1987, Scarlett et al. 1988). In brief, we routinely carry out 8-15 measurements on a sample from each patient/experimental animal. Generally, we consider the mean of these measurements as a single value, and base our statistical tests only on the number of patients/animals used, not on the total number of measurements. Student's t-test is used to distinguish between control and experimental/patients groups. In some cases we have resorted to analysis of variance.

Microprobe Analysis in Medicine - Present Practice and Future Trends

John D. Shelburne J. Allan Tucker
Victor L. Roggli Richard W. Linton+
Peter Ingram*

Department of Pathology
Duke University and V.A. Medical Centers
Durham, North Carolina 27710

This paper reviews current applications of microprobe analysis in medicine with emphasis on medical practice as opposed to research applications. In current medical practice, microprobe analysis is usually accomplished on an electron microscope (EM) equipped with an energy dispersive x-ray detector (EDX). Accordingly, electron probe microanalysis (EPMA) is emphasized in this communication. However, newer technologies such as ion and laser microscopes are now being used in a few laboratories for medical studies; therefore these will be briefly considered as well. For more extensive discussions of all of these topics, the reader is directed to our recent publication (Ingram et al., 1989).

Those medical conditions which are usefully studied by EPMA in order to provide diagnostic information include the pneumoconioses (asbestos, "hard metals", minerals, etc.), cystic fibrosis, renal stones, unexplained granulomas, and unexplained pigments or deposits (Shelburne, et al., 1989).

By "diagnostic" we mean that the information obtained can lead to an action that affects the particular patient in question in addition to the research implications. These actions are not solely medical but include legal, industrial and public health actions as well.

From an historical perspective, a major change in medical work is a steady move from qualitative microanalysis to quantitative microanalysis. For example, in the field of pulmonary particulates, the qualitative identification of quartz particles or asbestos

*Research Triangle Institute, Research Triangle Park, North Carolina 27709; +Department of Chemistry, University of North Carolina , Chapel Hill, North Carolina 27599-3290

fibers in the lung is of limited value, as almost all adults have some particles of this type in their lungs. The more important question is whether the numbers of these particles are increased both when expressed per lung or per gram of lung tissue, and focally as it relates to pathologic lesions. The second major trend is an improvement in specimen preparation, with a growing emphasis on the use of cryotechniques. The current approach of using chemically fixed tissues suffers from the obvious disadvantage of loss of ionic information. For many medical problems such as in renal pathology, the ionic information can be as important or more important than the usual structural information obtained with conventional techniques. Thus, the next major frontier in medical microprobe analysis probably will be the increasing use of quantitative x-ray microprobe analysis on cryosections for biopsies such as kidney and heart. The information generated from these studies will not only be of importance in understanding pathophysiology, but will also be of importance in normal human physiology. The third major trend is the increasing tendency to use new instruments which are capable of providing molecular information to supplement the elemental information already available from electron microprobe analysis. Laser microprobes and ion microprobes employing mass spectrometric detection appear promising for localizing small molecules such as dopamine and epinephrine to specific organelles. This is especially true when post-ionization approaches are employed to improve the ionization efficiency of the vaporized molecules. As currently configured, these instruments have already proven extremely valuable for detecting low levels (e.g. ppm) of trace elements such as aluminum in biological samples (Linton, et al 1989). Below we comment on some of these types of studies.

Qualitative Analysis

At the present time, most diagnostic cases are analyzed qualitatively using EPMA as applied to conventionally processed, routine histologic sections (Shelburne et al., 1989). Usually an energy dispersive X-ray detector (EDX) is used. Examples of this type of qualitative analysis would include establishing the diagnosis of zirconium granuloma, the identification of thorium deposits, and the identification of fluorine as a marker of Teflon deposits. In many cases the specific identification of foreign material has implications not only for the original patient, but also for many others, as identification may result in prevention. For example, Leong and colleagues (1982) used EPMA to identify particles of silicon in liver biopsies from patients with chronic renal failure treated by hemodialysis. These particles seemed to be related to hepatic inflammation and fibrosis, and could not be identified by light microscopy alone. Once EPMA had identified silicon in the inclusions, they traced the silicon to a segment of silicone tubing located in the roller pump of the dialysis machine. The vendor changed the manufacturing process, eliminating the problem. Cases such as these suggest to us that the qualitative analysis of conventionally fixed histologic sections can provide extremely valuable information.

Quantitative Analysis - Pneumoconioses

In terms of numbers of patients studied, the major current clinical application of EPMA is in the field of the pulmonary pneumoconioses, especially asbestosis. Quantitative microprobe analysis in diagnostic medicine began with asbestosis, in the sense that different types of fibers are now routinely identified in part by EPMA, and then accurately counted to determine number of fibers per unit of lung tissue (Abraham, 1989; Roggli, 1989).

In general, most patients studied have a history of exposure or biopsy pulmonary findings that are diagnostic of asbestosis. However, quantitation of the number and type of fibers adds significantly to the certainty of the diagnosis, and may point to a particular workplace as the site of major exposure.

Asbestos cases can roughly be grouped into three major types. Firstly, there are patients with a clear exposure history and unequivocal histologic findings at autopsy or on an open biopsy. In these patients, quantitative analysis of the fiber burden only confirms the diagnosis. More difficult cases include patients with pulmonary fibrosis secondary to other diseases such as polymyositis, and a history of asbestos exposure (Roggli 1989). Quantitation can assist in the determination of the importance of the asbestos exposure. Patients with mesothelioma may not have pulmonary asbestosis. Again, quantitation of the fiber burden can help in determining the possible etiologic role of asbestos exposure.

While considerable progress has been made regarding the techniques and procedures for the quantitation of asbestos fibers, the quantitation of other minerals in human lung tissue has been slow to develop. All adults have numerous inorganic particles in their lungs, including aluminum silicates, quartz, titanium oxides, etc. Therefore, in order to provide diagnostic information, it is often important that the particles be quantitated. That is, the numbers of particles of differing chemical composition needs to be sorted and counted. The chemical composition may be determined by qualitative EDX in simple cases (e.g., a silicon positive particle showing no other elements present in the EDX spectrum is assumed to be quartz); however, quantitative EDX may be needed to distinguish between, say, different clay minerals. Indeed, in many cases the qualitative EPMA analysis of this type of biopsy may be of little clinical value. For example, a biopsy report saying there are "many" aluminum silicates and quartz particles present in lung tissue is usually not very helpful to the clinician. Abraham (1989) has pioneered the in situ quantitation of particles in biopsy sections. Briefly, he uses conventional histologic sections mounted on carbon stubs, and examines them by backscattered electron imaging and EDX using a scanning electron microscope. His approach offers the advantage that he can correlate his quantitative data directly with histologic changes. We believe that there is a need for other laboratories to adopt this approach to the study of pneumoconioses, and to seek to develop agreement on procedures and measurements.

Quantitative Analysis - Cryotechniques

Even though physiologists and other investigators have developed excellent regimens for the quantitative analysis of electrolytes in normal and pathological tissues, the clinical application of these techniques is at present very limited (LeFurgey et al., 1988; Roomans, 1988). In view of the importance of ion regulation in physiology and pathology, it is our bias that probably the most exciting future applications of microprobe analysis will be in the study of electrolyte disorders, both at the cellular or light microscopic level, and at the subcellular or organelle level. An overview of some present and some very speculative future applications are given as Table I. Wroblewski and colleagues (Wroblewski and Edstrom, 1989) have pioneered the use of cryotechniques for the study of muscle biopsies. However, their approach is not widely practiced. Another exception is Roomans' (1989) use of EPMA to diagnose cystic fibrosis (CF) using fingernail clippings. Although cryo-techniques are not needed for these studies, the point is that tissues are not processed with aqueous chemical fixation and electrolytes are quantitated using EPMA. In research studies, Herlong et al (1988) have shown differences between CF patients and control patients using cryotechniques as applied to nasal scrapings. Quinton's (1978) studies of CF fluid samples have also revealed diagnostic differences. Thus one of the first clinical conditions which may usefully be studied for diagnosis with quantitative EPMA for electrolytes may be cystic fibrosis. The currently accepted standard for the diagnosis of cystic fibrosis, the pilocarpine iontophoresis sweat test, is difficult to perform. False positives and false negatives are frequent problems when inexperienced laboratories try to perform this test. In addition, even in excellent laboratories the distinction between positive and negative cases is not always clear-cut, as intermediate degrees of clinical severity and of salinity are known. It would be interesting to study these patients using cryo-techniques as applied to biopsy samples.

It is generally agreed that the field of diagnostic transmission electron microscopy began with kidney biopsies. We believe that a very important diagnostic application of quantitative EPMA would be for rapidly frozen kidney biopsies. There are a host of tubular disorders which are poorly understood and which could usefully be studied with EPMA. Indeed, extensive quantitative EPMA of normal human biopsies has not yet been accomplished. Current theories about the physiology of the kidney may have to be revised once hard data on electrolyte concentrations in different regions of the nephron are available; the study of pathological specimens will shed light on the normal physiology. Other types of biopsies which seem to lend themselves to this type of analysis include heart biopsies (myocardiopathies and acute myocarditis), and liver biopsies (detection of iron, copper, and possibly electrolytes in conditions such as Reye's syndrome).

A valid objection to the use of EPMA on biopsies is that unless the sample is frozen in a fraction of a second, the electrolyte changes caused by the biopsy procedure may obscure the true in vivo electrolyte distribution. While this objection is undeniable for many tissues and conditions, it is also possible that there are pathologic conditions that are so extreme that in comparison the changes caused by the biopsy procedure will be relatively unimportant. We are also confident that techniques for rapidly freezing biopsies will continue to improve steadily. High-pressure freezing techniques may be well suited for biopsy analysis because the depth of freezing extends several hundred microns.

Table I

A speculative overview of possible diagnostic applications of quantitative EPMA in non-pulmonary disorders.

Tissue	Pathology
Bone	Metabolic bone diseases (osteopenia) and metal deposition (aluminum)
Bone marrow	Anemias (Fe)
Ear	Hearing disorders (inner ear fluid, otoconia)
Eye	Corneal edema
	Glaucoma (giant vacuoles in endothelial cells of Schlemm's canal; aqueous humor)
Fingernails	Cystic Fibrosis
Heart	Myocardiopathies

	Myocarditis
	Hemochromatosis
Heart Valves	Distinguish different etiologies
Kidney	Drug toxicity (cyclosporine)
	Glomerulonephritis
	Metal toxicity (Pb, Bi, As)
	Tubular disorders (acidosis, necrosis)
Liver	Wilson's disease
	Hemochromatosis
	Primary biliary cirrhosis
	Primary sclerosing cholangitis
Muscle biopsies	Myopathies
Nasal mucosa	Cystic fibrosis
Nasal polyps	Secondary to cystic fibrosis vs. allergic
Skin	Psoriasis
Various	Oncology, cell injury (ischemia, effects of drugs or conditions such as bed rest and weightlessness on electrolytes)

In the general area of oncology there are numerous interesting questions awaiting microprobe study. These include the decision of whether certain low grade tumors are benign or malignant - for example, fibrotic disorders, adenoma vs. low grade hepatoma, papilloma vs. low grade carcinoma, gliosis, etc. Cameron et al (1989) have provided preliminary evidence that some malignant tumors are distinguishable from non-malignant parenchyma by quantitative EPMA. Differences in the secretions might also be demonstrable within gland lumens, e.g., atypical hyperplasia vs. carcinoma of the prostate or breast. Are there intracellular changes that correlate with invasion, that is, penetration of the basement membrane? If so, EPMA might be useful in the diagnosis of verrucous carcinoma (invasion is often impossible to identify histologically), borderline ovarian carcinomas and early cervical carcinomas. EPMA may be helpful in assessing differentiation. For example, the numerous dense core granules present in carcinoid tumor may show an unusual spectrum (e.g. high calcium).

In addition to the comments on electrolytes noted above in Table I, EPMA could theoretically contribute to treatment regimens in radiation therapy, being used as a guide to determine the effective radiation and/or hyperthermia "dosages" to tumors. The viability of tissues such as amputation stumps could be assessed. The use of EPMA to quantitate electrolytes in microdroplets of fluid obtained from sites such as kidney tubules, the eye or the inner ear is largely unexplored, but promising, as the techniques are now well established (Lechene and Warner, 1979; Roinel, 1988).

If microprobe analysis is such a useful tool in surgical pathology, why is it not used more frequently? There are many reasons for this including obvious considerations such as lack of available instrumentation in hospital settings, lack of personnel trained in the use of the instrumentation, and lack of knowledge; many pathologists have yet to hear

of this technique, and thus do not refer their appropriate samples for microprobe study. In Table II we have attempted to highlight a few of the major problems.

Table II

Major problems inhibiting the more widespread use of microprobe analysis in diagnosis

I.	**Knowledge of technique**	**Many physicians not aware of power/potential of technique. Application areas greatly underutilized include radiology (correlation of tissue location of contrast agents with images) and forensic medicine. "Missionary" work by cognoscenti needed.**
II.	**Personnel**	**Lack of courses, especially for specimen preparation/cryotechniques** **Lack of centers of excellence willing and able to accomodate students for training.**
III.	**Instrumentation**	
	A. Microscope	**Often not designed for microprobe analysis. Microprobe an "afterthought". Most microprobe systems designed for materials scientists, not biologists.**
	B. EDX	**Often not sensitive enough for trace element detection in toxicologic studies.** **Detection limit needs to increase roughly x 10.** **Slow analysis. Need optional simple systems for <u>speedy</u> qualitative analysis.**
IV.	**Specimen Preparation**	
	A. Cryofixation	**Cryofixation of a surface layer of ~10-15 microns is too shallow. Need ~100 microns of "ice crystal free" fixation.**
	B. Microtomy	**Techniques and equipment facilitating sectioning, section transfer, and freeze-drying could be improved. Still an intimidating technique.**
	C. Contamination	**Contamination with sulfur and silicon still difficult problems. Increased background from mass gain due to hydrocarbons and silicon can be difficult to avoid without column trapping and a cold stage.**
V.	**General**	**Lack of enough results. When important medical problems are "solved" by microprobe analysis, applications will increase.** **Lack of specificity of results. The electrolyte changes of non-specific cell injury may mimic the findings in a specific pathologic state.** **Lack of molecular information. Most work done with EDX to detect elements only. Many physicians more interested in molecular localization.**

A major point is that cryopreparation steps will have to become easier. Currently, there are only a handful of laboratories around the world that use proper cryopreparation techniques for quantitative EDX. Not only must techniques for rapid freezing improve to make it easier to freeze many microns of tissue, but also cryoultramicrotomy must improve. High-pressure freezing as described by Gilkey and Staehlin (1986) may be ideally suited for human biopsy samples, as ~500 microns of well-frozen tissue can be obtained (Menco, 1986). Cryoultramicrotomy should be easier with samples frozen in this manner. Low temperature embedding techniques are also promising, but not yet proven (Wroblewski, et al., 1988). Secondly, techniques for quantitating electrolytes in cryosections must become routine. Techniques for imaging cryosections need to improve so that the pathologist will know what he is examining. Energy dispersive x-ray detectors need to improve. For many problems in toxicology, energy dispersive x-ray microanalysis is simply not sensitive enough to be truly useful. One approach to improving this sensitivity would be to employ far more detectors. For example, if ten detectors, each with a 30 mm^2 Si(Li) crystal, were aimed at a sample in an SEM with a large chamber, this factor of 10 improvement would be very significant for toxicologic and clinical studies. However, such a system would obviously be cumbersome, and thus we believe that detectors using a totally different technology should be designed that will be more efficient. A detector that would completely "wrap around" the sample would greatly increase the efficiency of x-ray detection. Mercuric iodide detectors may offer future improvements (Iwanczyk, et al., 1984). Also, detection limits are better with x-ray fluorescence although spatial resolution is poor (Middleman and Geller, 1976).

In our work comparing energy dispersive x-ray analysis with SIMS and LAMMA, we have found that these latter two techniques generally do indeed provide better sensitivity. However, they are both destructive of the area of the sample being analyzed in contrast to electron beam instruments. Imaging is excellent at light microscopic magnifications in the usual laser microprobe, but at present one generally needs a thin section - for biology usually an Epon section, and that implies noncryotechniques. In addition, ion microprobes and laser microprobes are poor at quantitation because of wide variations in elemental sensitivity coupled with large matrix effects on ionization efficiences; the use of post-ionization techniques, such as with pulsed lasers, may significantly improve quantitation by reducing both of these problems. Also, field emission liquid metal ion sources are being developed that allow lateral resolutions approaching 20 nm. These will permit direct correlation of ion induced secondary electron images with ion induced secondary ion images, thus greatly improving current imaging with SIMS (Levi-Setti, et al., 1988). However, electron beam instruments currently offer the best chance to provide both high resolution imaging along with the ability to quantitate electrolytes in cryosections.

As noted at the end of Table II, the continued development of instruments such as LAMMA and SIMS and techniques such as cathodoluminescence is critical to the long range success of microprobe analysis in medicine, as these techniques can provide molecular information. Recent work by Hillenkamp (Karas and Hillenkamp, 1988) in detecting large organic molecules with time-of-flight (TOF) spectrometers offers the long-term prospect of major improvements in medical microprobe analysis. Detection limits in the subnanogram range of molecular masses exceeding 10,000 daltons appear to be feasible.

Conclusions

We conclude that microprobe analysis is no longer solely important for medical research - the findings often have diagnostic, therapeutic, and/or legal significance for the patient. During the next decade it will be important to broaden the scope of medical microprobe analysis from the current focus on qualitative EDX to the more widespread use of quantitative EDX, cryotechniques and techniques for molecular microprobe analysis.

References

Abraham JL, Burnett BR (1989) Quantitative in situ analysis of inorganic particulate burden in tissue sections - an update. Chapter 6 In: Ingram P, Shelburne JD, Roggli VL (eds) Microprobe analysis in medicine. Hemisphere Publishing Corporation, Washington, DC pp 111-131.

Cameron IL, Smith NKR (1989) Applications of electron probe x-ray microanalysis to the study of ionic regulation of growth in normal and cancer cells. Chapter 15 In: Ingram P, Shelburne JD, and Roggli VL (eds) op cit. pp 291-301

Gilkey JC, Staehlin LA (1986) Advances in ultrarapid freezing for the preservation of cellular ultrastructure. J Electr Micros Techn 3:177-210

Herlong JR, LeFurgey A, Ingram P, Mandel LJ, Hawkey LA (1988) Quantitative x-ray imaging of human cystic fibrosis nasal epithelium. In: Newbury DE (Ed) Microbeam Analysis. San Francisco Press, San Francisco, pp 447-450

Ingram P, Shelburne JD, Roggli VL (Editors) (1989) Microprobe analysis in medicine. Hemisphere Publishing Corporation, Washington, DC

Iwanczyk JS, Dabrowski AJ, Huth GC, Bradley JG, Conley JM, Albee AL (1984) First use of a mercuric iodide (HgI_2) energy dispersive x-ray detector in a scanning electron microscope. Scanning Electron Microscopy, 1:9-14

Karas M, Hillenkamp F (1988) Laser desorption ionization of proteins with molecular masses exceeding 10,000 daltons. Anal Chem 60: 2299-2301

LeFurgey A, Bond M, Ingram P (1988) Frontiers in electron probe microanalysis: application to cell physiology. Ultramicroscopy 24:185-220

Leong AS-Y, Disney APS, Gove DW (1982) Spallation and migration of silicone from blood-pump tubing in patients on hemodialysis. NEJM 306:135-140

Lechene C, Warner R (1979) Electron probe analysis of liquid droplets. In: Microbeam Analysis in Biology. Lechene C, Warner R (Eds) Academic Press, New York, pp 279-296

Levi-Setti R, Chabala JM, Wang YL, Hallegot P (1988) High-resolution ion probe imaging and analysis. In: Newbury DE (ed) Microbeam Analysis - 1988, San Francisco Press, San Francisco, pp 93-101

Linton RW, Musselman IH, Bryan SR (1989) Laser and ion microprobe mass spectrometry - applications to human tissues. Chapter 16 In: Ingram P, Shelburne JD, and Roggli VL (eds) op cit. pp 303-333

Menco BPM (1986) A survey of ultra-rapid cryofixation methods with particular emphasis on applications to freeze-fracturing, freeze-etching, and freeze-substitution. J Electr Microsc Techn 4:177-240

Middleman ML, Geller JD (1976) Trace element analysis using x-ray excitation with an energy dispersive spectrometer on a scanning electron microscope. Scanning Electron Microsc I:171-179

Quinton PM (1978) SEM-EDS x-ray analysis of fluids. Scanning Electron Microscopy 2:391-397

Roggli VL (1989) Scanning electron microscopic analysis of mineral fibers in human lung. Chapter 5 In: Ingram P, Shelburne JD, and Roggli VL (eds) op cit. pp 97-110

Roinel N (1988) Quantitative x-ray analysis of biological fluids: The microdroplet technique. J Electr Micros Techn 9:45-56

Roomans GM (1988) Quantitative x-ray microanalysis of biological specimens. J Electr Micros Techn 9:19-43

Roomans GM (1989) Cystic fibrosis. Chapter 9 In: Ingram P, Shelburne JD, and Roggli VL (eds) op cit. pp 163-175

Shelburne JD, Tucker JA, Roggli VL, Ingram P (1989) Overview of applications in medicine. Chapter 3 In: Ingram P, Shelburne JD, and Roggli VL (eds) op cit. pp 55-77

Wroblewski J, Wroblewski R, Roomans G (1988) Low temperature techniques for x-ray microanalysis in pathology: alternatives to cryoultramicrotomy. J Electr Micros Techn 9:83-98

Wroblewski R, Edstrom L (1989) Disorders of skeletal muscle. Chapter 12 In: Ingram P, Shelburne JD, and Roggli VL (eds) op cit. pp 219-236

Acknowledgement

The authors thank Marie Thomas for typing this manuscript.

Conference Discussion

J. Morgan: In many respects you, as a pathologist attempting to make accurate diagnoses on human material, are in the least favorable microanalytical position of all, because you must make your assessment on the basis of n=1 values. To compensate you expose that specimen to a spectrum of (correlative) observations using different analytical and imaging methods. However, you face another severe intrinsic problem: the specimen is presented to you in a form where its composition may have been severely compromised during, for example, quench fixation. How limiting are the preparative and sampling problems in reality?

J.D. Shelburne: With regard to preparative procedures, we feel that these are usually not limiting. In current medical practice, most studies are directed at the detection of insoluble particulates. For these conventional processing with formalin is adequate. Indeed, to quantitate asbestos fibers we remove (digest) the tissue, so that there is no concern about tissue preservation. However, if one is interested in soluble material such as electrolytes or metals such as arsenic, cryotechniques are necessary. In that case the most difficult problem is getting the sample from the patient to the cryogen quickly. Obviously, propane is not welcome in the operating room. Until safer/easier techniques for rapid freezing are developed, the analysis must usually be restricted to the cellular, not subcellular, level of analysis.

With regard to sampling problems, these are actually about the same as for any problem in biology. The main difference is that it is more difficult to get samples (especially control samples), so that one must usually patiently collect data over a several year period before there are enough patients and age/sex matched controls to publish.

H.K. Hagler: The recent advances in cryomicroscopy (cryofixation, cryosectioning, cryotransfer) make it easy to process fresh biopsy material for surgical pathology and have results within 30 min. I have helped process a liver biopsy from a patient suspected of having Reye's syndrome and the diagnosis was made based on the presence of enlarged mitochondria. X-ray analysis did not yield any additional information in this case because of the redistribution of elements during the removal of the tissue. A limitation which was not explicitly mentioned by you was availability of funding to do much of the groundwork necessary to develop diagnostic elemental analysis. Do you forsee improvement in the availability of funding or will it continue to be performed using limited local funds and as extra time permits?

J.D. Shelburne: Funding will be needed if diagnostic elemental analysis is to become a reality in the near future. Especially critical is research on easier/better methods of freezing biopsies. National funding agencies such as the N.I.H. in the U.S. should establish several centers of excellence in biological microprobe analysis, for example using the high voltage EM facilities as models. If given a reasonable assurance of long-term funding, these facilities could serve as the site for much of the needed developmental research. Courses in all aspects of biological microanalyis could also be held at these centers.

Subject Index

Page numbers refer only once to the same article or conference discussion section.

Springer Series in Biophysics

Editor: P. M. Bayley

Volume 1

Ehrenberg/Rigler/Gräslund/Nilsson (Eds.)

Structure, Dynamics and Function of Biomolecules

The First EBSA Workshop
A Marcus Wallenberg Symposium

1987. XXI, 302 pp. 115 figs. ISBN 3-540-17793-[illegible]

Volume 2

Bianconi/Congiu Castellano (Eds.)

Biophysics and Synchrotron Radiation

1987. XII, 353 pp. 192 figs. ISBN 3-540-18183-[illegible]

Volume 3

Aebi/Engel (Eds.)

Cytoskeletal and Extracellular Proteins

Structure, Interactions and Assembly

1989. XIV, 366 pp. 140 figs. ISBN 3-540-50662-7

Springer-Verlag Berlin
Heidelberg New York London
Paris Tokyo Hong Kong

MIX
Papier aus verantwortungsvollen Quellen
Paper from responsible sources
FSC® C105338

If you have any concerns about our products,
you can contact us on
ProductSafety@springernature.com

In case Publisher is established outside the EU,
the EU authorized representative is:
Springer Nature Customer Service Center GmbH
Europaplatz 3, 69115 Heidelberg, Germany

Printed by Libri Plureos GmbH
in Hamburg, Germany